RAND

Understanding the Air Force's Capability to Effectively Apply Advanced Distributed Simulation for Analysis

An Interim Report

Robert Kerchner, John Friel, Tom Lucas

Prepared for the
United States Air Force

Project AIR FORCE
1946 - 1996

PREFACE

This report presents our preliminary findings and observations on how the Air Force can more effectively apply Advanced Distributed Simulation (ADS) technologies for analysis. We discussed ADS with the analysis and ADS communities, and participated in several ADS efforts, including the Synthetic Theater of War Europe (STOW-E), a Ballistic Missile Defense Organization (BMDO) technical engineering demonstration (TED), and the Airborne Laser (ABL) Test Series 7. As a result, we have identified several advantages and challenges ADS presents analysts. This report reviews several general ADS analysis issues, as well as several specific points. The emphasis of the report is on the improvements that are required in ADS in order to allow credible analysis. Note also that while our emphasis is analysis, several of these suggested improvements relate in part, or even in their entirety, to training issues.

This work was done in the Model Improvement Study in the Plans and Operations Project of RAND's Project AIR FORCE. This project is sponsored by General Tom Case of AF/XOM. It should be of interest to combat analysts in all of the military departments and the Office of the Secretary of Defense.

PROJECT AIR FORCE

Project AIR FORCE, a division of RAND, is the Air Force federally funded research and development center (FFRDC) for studies and analyses. It provides the Air Force with independent analyses of policy alternatives affecting the development, employment, combat readiness, and support of current and future aerospace forces. Research is being performed in three programs: Strategy and Doctrine; Force Modernization and Employment; and Resource Management and System Acquisition.

In 1996, Project AIR FORCE is celebrating 50 years of service to the United States Air Force. Project AIR FORCE began in March 1946 as Project RAND at Douglas Aircraft Company, under contract to the Army Air Forces. Two years later, the project became the foundation of a new,

private nonprofit institution to improve public policy through research
and analysis for the public welfare and security of the United States—
what is known today as RAND.

CONTENTS

FIGURES

TABLES

SUMMARY

This report summarizes the major findings of our work to date on Advanced Distributed Simulation and Air Force Analysis, including the identification of the major advantages and challenges associated with using ADS for analysis, the major areas where improvements will be required to best realize the benefits of ADS, and our preliminary recommendations for actions.

ADVANTAGES FOR ANALYSIS WITH DISTRIBUTED ENVIRONMENTS

It is important to note that the advantages described below are in some sense potential or unproven in that they are not realized automatically from the use of ADS, but only when ADS is part of a carefully designed analytic plan.

- **Provide a better treatment of human elements when virtual/live participants are involved.** The behaviors that result are generally far more realistic and credible than those achievable using only constructive models. Because of this, the utilization of warfighters in human-in-the-loop (HIL) runs, via ADS, can significantly improve the quality of an analysis effort.

- **Provide a superior ability to present results.** Many individuals will better absorb and believe analysis results when they are presented with the visual displays, HIL simulators, etc., found in ADS facilities like the Theater Battle Arena (TBA) and the Theater Air Command and Control Simulation Facility (TACCSF). While analysts often neglect the presentation of findings, their job includes this facet of an analysis, and it is critical to the objective of having an analysis make a difference in the decisionmaking process.

- **Allow for parallel processing.** The increased computer power, in comparison with that available to most stand-alone constructive models, can enable the simulation of larger scenarios, and can also make it possible to simulate a given scenario at a higher level of detail.

- **Achieve faster model development.** Although unproven, another technical benefit may accrue from the increased reuse potential of ADS components and databases. Significant problems need to be overcome, but the potential is there for analysts to assemble the assets needed to simulate a scenario of interest, at an appropriate level of detail, with far less cost and effort than it takes to augment a constructive model to give it the needed capabilities.

CHALLENGES FOR ANALYSIS WITH DISTRIBUTED ENVIRONMENTS

We have also identified these major challenges associated with the use of ADS for analysis:

- **Size and increased complexity.** Those responsible for the design of exercises and the analysis of their results find it much harder to understand the assumptions and limitations in a distributed environment, in large part because the expertise has become as distributed as the simulation components themselves. Managing and scheduling ADS experiments is also far more complex than managing and scheduling experiments that utilize only stand-alone constructive models, especially when the experiments are distributed over multiple locations. Finally, reduced reliability is inevitable, both because these complexities invite error and oversight, and simply because the much larger amount of hardware and software involved means there is more to go wrong.

- **Virtual/live participants.** The inclusion of human-in-the-loop (HIL), while arguably the biggest advantage of ADS to analytic efforts, is not without its downside. Problems associated with the use of human participants include learning curves, gaming, small sample sizes, and nonreproducibility.

- **Other technical problems.** These include network bandwidths and exercise reliability. Problems caused by bandwidth limitations are not necessarily intrinsic to the use of ADS, in that future improvements in network capacity may provide more than adequate bandwidth, but we suspect that users' appetites will always grow as capacity expands.

IMPROVEMENTS REQUIRED

Our research indicates that a number of technical and organizational improvements must be undertaken to yield useful ADS capabilities for both analysis and training purposes. These include:

- **Knowledge base for ADS analysis.** The complexities of analysis in an ADS environment require new analysis strategies that are not now well understood. For example, techniques to utilize a combination of ADS experiments involving HIL and purely constructive simulations are quite immature, but are key to successful analysis with ADS. Areas for improvement include overall analysis strategies and experimental design techniques. Furthermore, a greater cross-training between the analysis, training, and testing communities will provide analysts greater insight into how, for example, to deal with human subjects, and to give testers and trainers the advantages of constructive modeling. In general, successful ADS analysis will require the development of analysis leaders knowledgeable in new ADS analysis techniques, traditional constructive analysis, and human experimental techniques. The leaders also need a realistic understanding of the timelines, costs, and organizational efforts associated with conducting an analysis that utilizes ADS.[1]

- **Simulator fidelity.** Improvements are needed in three broad areas:
  - Visual capabilities can be singled out as far and away the most significant shortcoming perceived by pilot participants in STOW-E. Improvements here will be important for most analysis involving HIL pilots, or for pilot training using simulators. Improvements are less urgent when the emphasis is on participants who normally operate in a workstation environment, such as controllers and intelligence analysts.
  - Tactical communication fidelity improvements to provide realistic communications will be especially important for analysis involving advanced warfighting concepts with critical timelines, such as theater missile defense (TMD) "SCUD-hunting."
  - A consistent terrain picture is needed for all ADS participants across live, virtual, and constructive models.

- **Computer generated forces limitations.** Improvements are needed to the computer generated forces (CGFs) used in ADS exercises involving HIL, and also in the purely constructive phases of an analysis, both for reliability and behavioral "realism." Reliability (resistance to the gross failure, or "crashing," of a CGF) is primarily an issue for ADS exercises only; behavioral realism is an issue for both ADS exercises and the constructive analysis phases of efforts that also involve ADS.

- **Viewing an ADS experiment (Stealths).** Improvements are needed to provide a single workstation with a variety of views and situation awareness assistance features for analysts, controllers, and others who wish to observe an ADS experiment. Such improvements are feasible from a technology and cost perspective.

- **Network reliability.** For analysis, especially when there is a strong reliance on scenario outcomes measures such as exchange ratios or win-loss results, system crashes can be catastrophic. If an analyst includes runs interrupted by network crashes, with the accompanying breakdown of remote site interactions, he will almost certainly bias outcome measures. Thus, current networking capabilities, including those involving the Defense Simulation Internet (DSI), need improvement to approach nearly 100 percent reliability.

- **DIS maturity.** Distributed Interactive Simulation (DIS) components are often not ready to perform with the kind of reliability needed for analytic efforts, and the standard itself is in need of extensions to better support analysis. Broadly speaking, DIS Protocol Data Units (PDUs) present "what"

---

[1]In this context, the initiatives outlined in *A New Vector* (Department of the Air Force, 1995) to provide improved career paths for M&S professionals are an excellent start toward meeting this need.

information; analysis also needs "why" information that is now typically internal to individual DIS components. The DIS standard needs provisions to facilitate capturing both the what and the why information in a form that can be readily accessed by analysts.

- **Complexity of exercise logistics.** ADS is currently too expensive for most analysis effort budgets, and too cumbersome to meet many decision dates. Improvements here will take the form of better procedures, both manual and automated, and ADS infrastructure support teams who have acquired more experience and expertise.

RECOMMENDATIONS

Our recommendations for actions that the Air Force should undertake, with respect to ADS for analytic purposes, fall into three broad categories: Air Force ADS investment strategies, automated tools and procedures for ADS management, and recommendations for individual exercises.

Air Force ADS Investment Strategies

Although improvements are needed, we see a great potential role for ADS/DIS in analysis, training, and mission rehearsal. Air Force requirements differ from those for the primary developers (U.S. Army and ARPA) of ADS technologies. To ensure that the Air Force gets the most from ADS technologies an Air Force investment strategy for ADS is needed. This will facilitate moving beyond the demonstration stage of ADS usage and into regular utilization for analysis and training purposes. The investment strategy will have to balance the potential benefits versus the costs and technological risks. A good starting point would be to prioritize the improvements required--including those listed in this report. Ideally, the priorities should be tied to their support for Air Force Program Objective Memorandum (POM) elements, and to programs to develop future systems, using a top-down approach like Strategies-to-Tasks.

Analysis efforts that will use ADS require specific research strategies that focus on the advantages ADS provides and also work around the weaknesses and problems, including the fact that ADS constitutes a highly constrained resource. This last point adds a major planning requirement: the analysis strategy cannot be developed ad hoc after the runs have been made. It also implies that the need for an ADS component in an analysis should be driven by the needs of the analysis team not simply by the fact that the component is available.

Automated Tools and Procedures for ADS Management

We have noted that the effort spent on ADS exercises is great. It is also rather error-prone, as can be expected when a technology is so new and when many individuals need to cooperate in a complex and nonroutine environment. Automated tools and standardized procedures (manual as well as automated) can be of great benefit in reducing both effort and errors. Automated tools would be extremely beneficial in assisting the distribution of databases and software upgrades to distributed components. As well as assist in the actual distribution and installation, such tools would ensure (at exercise initialization) that the proper versions are in fact being used.

Another candidate for procedure development involves testing the behavior of computer generated forces (CGFs) and other simulation components. Such testing does not appear to be a good candidate for full automation, because this validation phase will generally have unique features for each exercise. However, procedures and partial automation should be helpful.

These procedures and automated tools are not intrinsically Air Force specific, although the validation-oriented tools will likely benefit from testing oriented toward Air Force systems and missions. However, we believe that the DIS community as a whole has been somewhat slow to recognize that many of the "awkward" encounters in current DIS exercises are not one-time occurrences that one "just works through." Rather they are symptomatic of systemic problems that are likely to occur again and again. Someone needs to step up to the challenge of mitigating these problems; if undertaken by the Air Force then Air Force interests will more assuredly be considered.

Recommendations for Each Exercise

The following recommendations can be immediately implemented for all exercises that include analysis objectives:

1. **Define realistic exercise objectives.** Getting credible analysis out of ADS exercises requires realistic objectives that reflect the limited number of runs available with HIL, the fidelity of the models involved, and the maturity of ADS as a whole.

2. **Improve management of exercises.** The management of ADS exercises is extremely difficult, with joint participants at multiple sites. Some measures to enhance the quality of the exercises (all at some cost) are:

- Explicitly and rigorously test the components for adherence to DIS standards.
- Routinely plan tests of the experimental setups.
- Place a hold on software and database modifications at some point prior to the exercise.
- Develop and use a set of predefined guidelines for handling hardware and software failures.

3. **Implement timestamps for all Air Force DIS activities.** Timestamps should be immediately added to all DIS PDUs. Time coordination among the sites can most easily be accomplished using GPS signals, but other alternatives, such as synchronizing over a voice telephone line, may be sufficiently accurate for many applications.

4. **Develop summary documentation of all models.** Many of the models used in DIS exercises were developed for demonstration purposes and have little or no documentation of their assumptions and limitations. Such documentation is invaluable for assessing the adequacy of an exercise configuration with respect to an analytic approach.

ACKNOWLEDGMENTS

We base this work on observations we made at several ADS exercises, as well as numerous meetings with many people in the vanguard of ADS analysis efforts. Each ADS effort invariably brings some new insights, so many sources contributed to the ideas within this report. At the risk of missing someone, we list the following people who made particularly important contributions. Colonel Ed Crowder of Air Force Studies and Analyses Agency (AFSAA) provided valuable feedback to earlier briefings and connected us with several ongoing efforts. Lieutenant Colonel J. D. Dennison of XOMW facilitated our participation in the Synthetic Theater of War Europe (STOW-E) and shared with us the details regarding the logistics required to make the STOW-E become a reality. Colonel Ron Stanfill, also of XOMW, stimulated our thinking on relating suggested improvements to investment strategies. Dr. Herb Bell at Armstrong Laboratories help formulate our thinking on assessing the training benefits of ADS exercises. George Degovanni of Booz-Allen Hamilton showed us Warbreaker's approaches to using ADS for analysis. Throughout this project we have been aided by several Theater Battle Arena (TBA) personnel, including Lieutenant Colonels Brian Millburn, Frank Zawada, and Kevin Martin.

Many thanks are also due to the gracious people at the Theater Air Command and Control Simulation Facility (TACCSF) who hosted several visits there, especially Lieutenant Colonel Denny Lester and Captain Tom Akin. Lieutenant Colonel Milt Johnson, at the National Test Facility (NTF) provided insights into the Ballistic Missile Defense Organization (BMDO) theater missile defense (TMD) distributed technical engineering demonstration (TED). Finally, reviews by Louis R. Moore and Steven Bankes of RAND have greatly improved both the presentation and content of this report.

GLOSSARY

ABL	Airborne laser
ACC/XP-SAS	Air Combat, Deputy Chief of Staff, Plans and Programs, Studies and Analysis Squadron
ADS	Advanced Distributed Simulation
AFA	Air Force Association
AFSAA	Air Force Studies Analyses Agency
AF/XOMW	Air Force Directorate of Modeling and Simulating Warfighting Support
AGs	Application gateways
ALO	Air liaison Officer
ALSP	Aggregate Level Simulation Protocol
ARPA	Advanced Research Projects Agency
A2ATD	Anti-armor Advanced Technical Demonstration
AWACS	Airborne Warning and Control System
BMDO	Ballistic Missile Defense Organization
CAS	Close air support
CGFs	Computer Generated Forces
CONOPS	Concepts of operation
DIS	Distributed Interactive Simulation
DSI	Defense Simulation Internet
EM	Electromagnetic
FOV	Field of view
HAE	High Altitude and Endurance
HAV	High altitude and endurance
HIL	Human-in-the-Loop
IDA	Institute for Defense Analysis
I/ITSEC	Interservice Industry Training Systems and Education Conference
LOS	Line-of-Sight
M&S	Modeling and Simulation
MOE	Measures of effectiveness
NCCOSC	Naval Command, Control and Ocean Surveillance Center

NRaD	Naval Command, Control and Ocean Surveillance Center, RDT&E Division
NTF	National Test Facility
PDUs	Protocol data units
PK	Probability of kill
POM	Program Objective Memorandum
RWR	Radar warning receiver
SA	Situation awareness
SAFORS	Semi-automated Forces
SEAF	Simulation Evaluation and Analysis Facility
STOW-E	Synthetic Theater of War-Europe
TACCSF	Theater Air Command and Control Simulation Facility
TBA	Theater Battle Arena
TBM	Theater ballistic missile
TED	Technical Engineering Demonstration
TMD	Theater Missile Defense
TOC	Transfer-of-control
UAV	Unmanned Aerial Vehicle
VV&A	Verification, validation, and accreditation
WAN	Wide area net

1. INTRODUCTION

One of the many significant challenges of developing and applying Advanced Distributed Simulation/Distributed Interactive Simulation (ADS/DIS) technologies centers on how ADS can be used to support analysis. This study is intended to improve the utility of ADS/DIS for the Air Force as a whole and for the analytic community within the Air Force in particular. This report summarizes our progress to date.

After a brief introduction and a summary of our approach, we present our observations of some recent ADS efforts--including the Synthetic Theater of War-Europe (STOW-E) exercise and several other activities--and our general conclusions about the use of ADS for analysis. We discuss the inherent advantages and challenges of ADS for analysis and describe an analysis approach that utilizes ADS in conjunction with constructive models. Finally we set forth our preliminary conclusions and recommendations about the role of ADS in Air Force analysis efforts.

WHAT IS ADS?

There does not yet appear to be a universally accepted definition of ADS. In this report we take a broad perspective based on the January 1993 report of the Defense Science Board Task Force on Simulation, Readiness and Prototyping. Consistent with this perspective, when we refer to ADS we mean: The ADVANCED enabling technologies[1] that allow geographically DISTRIBUTED sites to share a "synthetic battlefield" with a mix of live, virtual, and constructive SIMULATIONS. *Live simulations* are operations with instrumented operational equipment--such as the aircraft in Red Flag or the tanks at the National Training Center (NTC). *Virtual simulations* involve humans in simulators--such as aircraft cockpit simulators. *Constructive simulations* are computer models.[2] The

---

[1]Such as networks, architectures, formal standards, and protocols.

[2]Some definitions of *constructive simulation* include war games, where humans dynamically interact with the models--typically for command type decisions. We would tend to call these models virtual, or an aggregate consisting of virtual and constructive components. Our use of

ADS synthetic battlefield can be used for training, analysis, prototyping, etc.

Another, older, term for these technologies and vision is Distributed Interactive Simulation (DIS). However, DIS is now used to refer specifically to ADS in the context of a specific set of standards and protocols, including IEEE 1278. Thus, the term ADS is broader than DIS because it includes distributed simulations such as the Aggregate Level Simulation Protocol (ALSP) confederations, which do not conform to DIS standards.

A PRAGMATIC MOTIVATION FOR STUDYING ADS

This report details technical reasons why ADS can be useful to Air Force analytic efforts. In addition to these reasons, there is a pragmatic argument for increasing the Air Force's interest in ADS, and for helping to guide the evolution of DIS standards to support Air Force needs in general, and analysis needs in particular.

Modeling and Simulation Resources Go To ADS

ADS technologies are receiving an increasing share of DoD M&S resources. *The DIS Vision* (DIS Steering Committee, 1993) states that "Almost every major simulation being procured today will become a part of ADS." The services use simulations extensively today to equip, train, and employ our forces; the ADS standards and protocols (i.e., the enabling technologies) that evolve from these simulations will have a great influence on our military capabilities.

Training and Technology Currently Drive DIS Evolution

The genesis of ADS technologies was the ARPA-sponsored SIMNET (or Simulator Networking) program for Army armored combat training. The

---

the term "virtual" is not quite as broad as the term "open simulation," in that it excludes the "open" case where the only human interaction occurs while the simulation pauses for manual inputs, generally to direct behaviors for the next time period. Unless otherwise stated, when we refer to constructive models we mean a "closed simulation," i.e., no human participation. Occasionally our discussion of constructive models includes semi-autonomous forces, where humans interact not as warfighters, but rather to correct for shortcomings of otherwise fully automated simulation components.

ARPA and Army communities are still the driving force behind ADS evolution. Most of the people defining the new standards and protocols are from these communities. As an illustration of the comparative underrepresentation of the Air Force, less than 4 percent of attendees at the March 1995 DIS Workshop chose to attend a meeting for those with Air Force interest. If the Air Force does not become more involved in shaping the DIS evolution, new capabilities will not fully support Air Force needs.

The Analysis Community Has an Interest in Influencing Future DIS Developments

While training needs are driving the evolution of ADS, much of its promise relates to analysis issues. The Defense Science Board (January 1993) report states that ADS can provide the means to "transform the acquisition process from within." While ADS provides some significant analytic advantages over traditional methods, many of its features make it more difficult to use effectively. Recent experiences with the initial efforts to use these technologies for analysis--such as the Anti-armor Advanced Technical Demonstration (A2ATD) and the Ballistic Missile Defense Organization Technical Engineering Demonstration (BMDO TED)--illustrate how challenging performing analysis with distributed and interactive simulations can be. Many people in the analysis community are concerned that ADS is viewed as a panacea and that there is a risk that it will be misapplied as a tool in procurement and employment decisions. For example, using ADS in operational effectiveness evaluations, without understanding the limitations of the constituent models or the synthesized whole, could contribute to expensive procurement errors. This concern is magnified for Air Force systems because the Air Force has invested less than the Army in DIS technology, and Air Force systems have some unique requirements.

DIS is a radically new technology and some reflective thought on how it can best be used for analysis is warranted.

STUDY OBJECTIVES AND PRODUCTS

Primary Study Objective

The primary objective of this study is to improve the utility of ADS for Air Force analysis applications. Analytic applications of ADS are emphasized because of the increased interest in using ADS for analytical purposes (effectiveness analysis, employment strategies, trade studies, virtual prototyping, etc.). Before credible Air Force analysis with these technologies can become feasible, many challenges need to be addressed. These challenges are not necessarily relevant to (or perceived as relevant by) the technology and training communities leading the ADS evolution process.

Focus on Distributed Interactive Simulation (DIS)

As noted earlier, Air Force interest in distributed simulation for analytical purposes extends beyond the DIS infrastructure, that is, beyond the product of an ongoing series of workshops that are creating a set of formal standards to support ADS. As a practical matter however, most of the current development is with DIS. Thus, we have focused our efforts there, but at the same time have tried to retain a more general perspective.

Desired Study Products

We will produce concrete products from this study, including (1) usable guidelines to help analysts who are considering the use of distributed (constructive) and/or interactive (virtual) simulation; (2) the identification of tools that improve the ability to perform analysis in a distributed environment; and (3) specific recommendations for where investment is needed to improve Air Force capabilities to use ADS for analysis. The foundation for obtaining such products is understanding what the advantages and challenges of ADS-supported analysis are. Only then, in the context of a comprehensive research plan for an analysis, can one address whether an analysis should utilize distributed and/or interactive simulation. We believe that when ADS is credibly used to support an analysis effort, it will normally be integrated into an analytic plan where it supplements and complements other analysis tools.

2. APPROACH

Our study approach had three components. The first was to investigate several distributed analysis efforts around the community. Next, to obtain in-depth understanding we participated in several distributed exercises. Finally, informed through this investigation and participation, we are developing strategies to support ADS analysis.

INVESTIGATION OF DISTRIBUTED ANALYSIS EFFORTS AND RESEARCH

Literature Review

Our literature review was broad, ranging from abstract theoretical papers to detailed study reports from academia, industry, and the services. We reviewed high-level "vision documents," such as the DIS Steering Committee's *The DIS Vision* (1993) and the Air Force's *A New Vector* (1995), as well as some early ADS-supported analysis reports, such as TRADOC Analysis Center's *Results of the M1A2 SIMNET-D Synthetic Environment Post-Experiment Analysis* (1993). The bibliography cites additional relevant reports. We also examined how classical experimental design techniques can be applied or extended to ADS experiments.

Survey of Air Force Capabilities and Current Efforts

Much of our initial effort went into meeting with people at the forefront of DIS-based analysis efforts and demonstrations. We emphasized investigating Air Force capabilities and plans. Air Force facilities we visited include the Pentagon's Theater Battle Arena (TBA), the Theater Air Command and Control Simulation Facility (TACCSF) in Albuquerque, and the Simulation Evaluation and Analysis Facility (SEAF) in Grafenwoehr, Germany.

Efforts involving Air Force participation included STOW-E, the BMDO TED, and the Airborne Laser (ABL) Test Series 7. The results of our examination of and participation in these efforts are presented in Section 4.

Survey of Non-Air Force Efforts

Because the Advanced Research Projects Agency (ARPA) and the Army have sponsored much ADS development, they have more mature analysis efforts. We met with the Warbreaker and A2ATD programs (more on this in Section 4) to learn what they have discovered in their "rubber hits the road" efforts. We also visited the National Test Facility (NTF) in Colorado Springs and participated in the semi-annual DIS Interoperability Working Groups. These working groups are defining the new architectures, standards, and protocols.

Identification of Distributed Analysis Techniques in the Community

RAND's unique role is to identify, generalize, and disseminate information about the difficulties encountered and the techniques developed to mitigate those difficulties. The community as a whole can thus benefit from advances made by the various efforts.

PARTICIPATION IN DISTRIBUTED ANALYSIS EFFORTS

Concern: In-depth Understanding of Problems Requires Hands-on Involvement

Actual participation in ADS-supported analysis efforts is essential to making sure that results produced will be of more than academic interest. The in-depth understanding of the subtleties of problems that comes from actually sitting down and trying to work through them is the major safeguard against promulgating naive or overly simplistic solutions.

Solution: Joined ADS Projects, Took on Specific Responsibilities

Our participation in the STOW-E, a BMDO TED, and an analysis effort sponsored by the Airborne Laser (ABL) program office showed us that many systemic problems are not clearly recognized as such by those who work on them. Although why this is so is not clear, we speculate that in the rush to reach an immediate goal, problems tend to be treated as one-time occurrences. They are either not identified as being caused by systemic defects that could be removed, or the effort of addressing the systemic issue is viewed as too costly or too time-consuming to be undertaken in the context of the current effort. The unfortunate long-term effect is that new manifestations of the systemic problem continue to appear. Example areas include problems with accommodating to requirements changes, and difficulties in reliably distributing updates.

Participation Focused on Analysis

Assisting in exercise design and some post-exercise analysis--mostly data extraction--has been emphasized. Our findings will be discussed in more detail later.

DEVELOPMENT OF STRATEGIES TO MITIGATE ADS SHORTCOMINGS

Develop Methods and Procedures to Maximize the Benefits While Mitigating the Difficulties

There are both new and exacerbated difficulties when an analysis includes distributed and/or interactive simulation. One of our key objectives is to identify general techniques that can mitigate some of these problems. Techniques we have identified are discussed in detail later in this report.

Promulgate Methods and Procedures in an "ADS Analyst's Guide"

To make our findings most useful to the analytic community, we intend to publish an "ADS Analyst's Guide" that will assist those who want to determine if ADS should play a role in their analysis. When ADS will be a component of their study, the Guide will assist analysts in all phases of their effort, including exercise and experimental design, exercise execution, and analysis. Its objectives are to provide both positive and negative support to analysts. Positive support includes broad analytic strategies for best utilizing different tools, such as virtual (human-in-the-loop) ADS exercises and constructive models (the latter may or may not take the form of a distributed simulation). "Negative" support includes warnings of problems that can be anticipated and techniques to mitigate them.

The state-of-the-art of analysis using ADS is still in its infancy. Accordingly, we see the ADS Analyst's Guide as a dynamic document that should be revised periodically to reflect the community's increasing understanding of this art. Our FY1996 effort constitutes an important start.

3. OBSERVATIONS FROM STOW-E

This section and Section 4 set forth our observations from reviewing and participating in several distributed efforts. We begin with a detailed discussion of the largest distributed effort to date, and the source of the majority of our insights--the Synthetic Theater of War-Europe (STOW-E). We then look at eight classes of suggested improvements that are largely based on our STOW-E observations.

STOW-E OVERVIEW

The STOW-E, an ARPA project, is the largest DIS exercise undertaken to date. It represented a portion of the Atlantic Resolve (formerly called Return of Forces to Germany (REFORGER)) training exercise. At a peak level there were 1860 joint entities playing in the STOW, all linked using DIS protocols. The STOW-E was strictly a technology demonstration effort; in fact, it was not "linked" to the overall Atlantic Resolve exercise (that is, neither system was aware of, or used the results of the other). However, the STOW-E missions were representative of those in Atlantic Resolve. The Army and Navy had constructive, virtual, and live elements in the STOW-E, while the Air Force had constructive and virtual participants. The virtual Air Force elements consisted of a variety of simulators and constructive models running in both Europe and the United States.

Air Force pilots flew a variety of simulator missions, including close air support, counterair, and interdiction--some joint with Navy aircraft simulations. Impressive aspects included a virtual Air Force aircraft in Grafenwoehr, Germany, which flew escort for a live Navy bombing aircraft at Cherry Point, North Carolina.

This was truly a global effort, involving over one dozen sites. There were virtual Air Force participants in Grafenwoehr, Lakenheath, TACCSF, TBA, and Armstrong Labs. Efforts are under way for more ambitious future exercises: STOW-97 is planning on up to 50,000 entities playing at once; the STOW-2000 goal is to have up to 100,000 entities playing at once.

RAND took part in assessing the training and mission rehearsal potential for the Air Force, with two representatives in Grafenwoehr and one at the TACCSF. Being on the inside provided us access to issues that also relate directly to analysis potential.

OVERALL IMPRESSIONS OF STOW-E

Major Technical Accomplishment

There is no question that STOW-E was a major success as a technical demonstration of the enabling technologies. A large number of sites in both the United States and Europe were linked with "reasonably good" network reliability. Many management issues, both low- and high-tech, were addressed, some successfully, and others well enough to enable a complex system to function. DIS interoperability was demonstrated across multiple sites, in the live/virtual/constructive domain, and also in the joint domain. None of these accomplishments is trivial, and none should be undervalued.

Technical Demonstration Only

Concluding that STOW-E was a demonstration of *actual* training, however, rather than a demonstration of technology and the potential for training and mission rehearsal, would be an error. Many improvements are still needed before participants--in particular Air Force operators --can have a positive training experience in this sort of exercise.

Although we emphasize that solutions to many current shortcomings appear feasible, not all of the known problems are being addressed at present, and we are skeptical about solving them in time for STOW-97 to demonstrate real Air Force training. In other words, we doubt that participants in simulations of the quality anticipated for STOW-97 will materially improve their readiness. It will be even more difficult for the next few STOWs to provide more than limited analysis. For example, fairly simple human factors analysis (such as measuring the timelines associated with an Airborne Warning and Control System (AWACS) operator) requires recording information internal to simulation components (such as when tracks are first established and displayed). It is not necessarily difficult, but it is necessary to arrange to capture such information in advance, and this is unlikely to occur absent a specific analytic objective.

Potential for Training and Mission Rehearsal

The potential of STOW-scale exercises for training and even mission rehearsal is clear, and is very large. The training advantages that STOW technologies may bring include the ability to train in the context of large scenarios, to train without range safety and emissions constraints, to participate in joint training, and to allow colleagues who are geographically distant to work together in advance of actual combat or live training exercises. *These benefits are difficult or impossible to achieve otherwise,* for both technical and fiscal reasons, and their value makes the STOW vision worthwhile. We noted the potential benefits of ADS for training and mission rehearsal previously, but this technical demonstration went a long way toward illustrating this previously abstract vision in a large-scale joint scenario.

Potential for Analysis Using Results Of STOW-Sized Exercises

We also considered the value that STOW-E type exercises might provide for analysts. For example, such exercises could be a source of useful human factors data which could be used to inform constructive models (such as operator timelines).[1] More generally, the exercises could be a source of engagements whose "flavor" could be a source of insight into the combat interactions exhibited by the scenarios being analyzed. This flavor should also be reproduced by constructive models.

(We realize that the meaning of the above statement is subjective since flavor is not rigorously defined: It has to do with recognizing common features of the behaviors observed in sets of runs, but just which features are important depends on many specifics of each scenario. What we desire is a means for assessing the credibility of the kinds of engagements that occur in the model vis-à-vis the engagements observed

---

[1]However, in most cases we would expect that better human factors information could be obtained using virtual simulators in much smaller, and carefully controlled, experiments. One case where STOW-size exercises might provide better data would be where the participants being measured are directly influenced by the size of the scenario, perhaps AWACS operators faced with a very complex air picture.

in the exercise.[2] This issue is particularly relevant in validating the constructive models.)

When significant inconsistencies are apparent between virtual and constructive runs, they provide an important opportunity to investigate differences and resolve errors. This opportunity can be used to enhance understanding of the simulations, including the ability to articulate the causes of the differences. This effort can illuminate important assumptions that might otherwise remain obscure. When the constructive models are judged consistent with the virtual simulations, they can be run extensively to provide statistical power and sensitivity analysis.

IMPROVEMENTS REQUIRED

We now look at a variety of potential improvements that would increase the training and analytic utility of a STOW or similar exercise. Our discussion occasionally detours from the overall analysis emphasis of this report and notes improvements needed for training.

Improvements Required for Simulator Fidelity

Visual. For Air Force training, the most significant shortcoming perceived by pilot participants concerned deficiencies in the visual capabilities of simulators. This is true of many stand-alone simulators too, but large DIS exercises imply full mission simulation and necessitate awareness of the pitfalls associated with interfacing a part-task trainer into the DIS environment. One difficulty with visuals lies in the lack of field of view (FOV) which forces simulator pilots to deviate from usual tactics to complete their close air support (CAS) missions. Such deviation is not "training like we fight" and may in fact be negative training. Another visual display deficiency is a lack of the resolution needed to support realistic target acquisition.

Tactical Communications. Tactical communication fidelity was next on the participants' list of shortcomings. The aircraft simulators at Grafenwoehr and TACCSF, for instance, had no ability to support

---

[2]We leave this point unresolved and rely on application specific domain experts to resolve whether or not two sets of engagements are consistent in their underlying mechanisms, even though, perhaps, the detailed evolution of the engagements differ.

realistic tactical communications. In many cases participants relied on commercial communications, and at one site, simulator operators often shouted to nearby operators rather than use a simulated communications network.

Terrain Correlation. Terrain correlation is a continuing problem because different participants have differing perceptions of terrain height. As a result, tanks can appear to fly through the air, and air-to-ground attackers can find themselves trying to acquire underground targets. This is in part due to database discrepancies,[3] but it is also associated with the use of different algorithms to process and display the terrain data. The adaptation of a single algorithm standard is not a feasible fix to the latter problem, because different types of simulators have differing needs. For example, the image generating requirements and line-of-sight (LOS) calculations for a fast moving aircraft at elevation are inherently different than those for a relatively slow moving ground vehicle.

Hardware Models. Many hardware models used in STOW-E--such as those for radar, infrared sensors, and weapons effects--were so oversimplified that they interfered with training. For instance, the F-16C radar in the Falcon Star simulator had the capability to acquire and lock onto ground targets, something the pilot assured us never happened in the actual aircraft. Other hardware models, like those for radar warning receivers (RWRs), were irrelevant because there was no exercise support for emissions protocol data units (PDUs). The resulting lack of missile warning was considered a major shortcoming. Taken together, these deficiencies resulted in pilots using different systems and tactics to achieve their mission objectives--again a potential source of negative training and erroneous analytic insights.

---

[3]Database discrepancies are superficially easy to eliminate. This is "conventional wisdom." However, our experience with the distribution of software and databases to a distributed community leads to the conclusion that this is a non-trivial problem that requires imaginative approaches to its solution. One possible means of addressing data consistency problems in the DIS environment is to make use of the connectivity implied by the network to (1) distribute the data and (2) implement run time (exercise initialization) checks on the versions of databases and software being used by exercise components.

From an analysis perspective, simulator fidelity shortcomings will bias outcomes. For instance, the lack of a full FOV may handicap virtual combatants when engaging constructive opponents. Of course, many other elements may systematically favor one side or the other when entities from different models and different types of participants (i.e., live/virtual versus constructive) are used. Analysis efforts will be required to account for these biases--lest they produce erroneous conclusions.

Improvements Required to Computer-Generated Forces

While ADS technologies facilitate human participation in simulations, for large STOWs the overwhelming majority of participants are computer generated forces (CGFs). This will become even more true as the STOWs scale up by more than an order of magnitude in the next few years. Thus, CGF performance will greatly influence the utility of large STOW exercises. We see this reason alone as sufficient to justify serious attempts to improve CGFs.

Reliability. Simulation application reliability, and CGF reliability in particular, was the weak link in STOW-E, more so than Defense Simulation Internet (DSI) reliability. Computers that hosted the CGFs crashed several times each day. At several sites this occurred in large part because STOW-E offered the first opportunity to stress some applications with very large scenario sizes. One might thus tend to treat this as a temporary problem that does not require special attention. However, we believe that the component applications will never be static, but will be frequently modified in response to changing user requirements. It seems inevitable that bugs introduced by these modifications will continue to crop up in actual exercises.

The problems caused by large numbers of participants, mentioned above, constitute one class of unanticipated situations that caused CGF failures. Other unanticipated situations, often associated with unusual scenarios, will occur and can be expected to cause problems resulting in CGF failure. Comprehensive preventive measures are difficult to envision, but an awareness of the assumptions that underlie the scenarios used (perhaps unconsciously) for CGF design can go a long way toward improved robustness. For example, the rules in an expert system typically do not include explicit tests against items assumed valid for the entire scenario. Such rules fail when the scenarios in fact depart from the assumed regime. In many cases, dependencies on hidden assumptions can be avoided by a more careful design and development process.

Behavior. Another problem we saw as we followed selected engagements in STOW-E was that some constructive entities seemed nonreactive. An example of nonreactive behavior occurred when a ground-based observer destroyed CGF tanks to provide a visual marker to help direct in a virtual CAS bombing run. As the ground vehicles were being attacked by the ground observer, they did not react (i.e., did not run, return fire, etc.) in response. Identifying and correcting behavior like this is important if CGF behavior acceptable for training and analysis uses is to be achieved. A concern in a distributed environment is that CGFs at different sites will not be "balanced."[4] That is, they are modeled so that one site has a distinct advantage over the other (or over virtual entities). For example, some CGFs ignore cultural terrain features when computing line-of-sight (LOS) while others do not, and some CGF entities use perfect state knowledge of other players when making decisions while others model limited situation awareness.[5] These differences may result in participants taking erroneous tactical lessons from their training or analysts drawing incorrect conclusions. One potential approach to address this asymmetry would be to do extensive pairwise testing (and modifications) between the distributed sites and models, thereby ensuring pairwise "balance."

With very few runs available from an analysis perspective, CGF reliability becomes a paramount concern, exacerbated by the relatively small time windows within which these exercises take place. It may only be during these times that distributed CGFs interact in a way that

---

[4]This is sometimes referred to as a fair fight.

[5]We have not explicitly looked at the CGFs used in STOW-E for these features; they are presented as examples of constructive model limitations that are quite common. In our experience, virtually all models of simulated decisionmakers compromise on the modeling of realistic situation awareness to a greater or lesser degree.

problems are manifested. Experience with recently extended large
constructive simulations shows us that a significant amount of time is
spent ironing out such problems, time that will not typically be
available for tightly scheduled distributed exercises. In fact, many
analysts plan an iterative strategy to cope with this difficulty.

Another problem associated with CGF reliability stems from the
problem of how to recover an exercise from a CGF "crash." Do you bring
the CGF back up as close to the condition that it was in when it
crashed? Do you recycle the CGF (and hence by implication the entire
exercise)? What of the live aircraft systems that may not have the fuel
available to recycle to scenario start? All of these issues must be
resolved early in the planning stage of large STOW-sized exercises, if
their analytic potential is to be realized.

Automated CGFs. CGF behavior realism is particularly difficult to
achieve in the complex air combat environment. We are concerned that
the ModSAF technology, which threatens to become the standard for DIS
CGFs, may be inadequate to the task. ModSAF's strategy of using *semi-
automated* forces relies on human operators to oversee a number of
automated participants, and to "correct" for deficiencies in their
behavior. However, this brings with it the unrealistic situation
awareness typically possessed by SAF operators. At best, it must result
in a set of automated players who share a single perceived reality (the
operator's) and who thus benefit by an unrealistic ability to avoid
confusion. These deficiencies are likely to be particularly important
in air combat scenarios, where surprise and confusion play a dominant
role in determining outcomes.

An equally compelling reason for wanting CGFs to be fully automated
is a corollary of the requirement for future STOWs to include very large
numbers of entities. If there are to be 100,000 entities involved,
upward of 90 percent of them will be CGFs. Even if a semi-automated
forces operator could control 100 entities (extremely unlikely, in our
opinion), this would still imply the need for 900 SAF operators; several
thousand such human operators is a more likely number. This becomes
expensive, and the issue of training so many operators becomes important
as well.

Improvements Required for Exercise Discipline

Exercise discipline is required for training and analysis. We emphasize that we do not intend to criticize any lack of exercise discipline in STOW-E. Because STOW-E was a technical demonstration, not actual training, exercise discipline was not a priority. Exercise discipline issues should be viewed as considerations for future training or analysis exercises.

There were, however, several doctrinal violations and artificial assistance by operators and assistants in STOW-E that could interfere with actual training or analysis efforts. These included doctrinal violations, and artificial actions to "make the mission work." Doctrinal violations include multiple go-arounds in air-to-ground attacks; artificial actions include helping the pilot locate targets on the simulator and the Air Liaison Officer (ALO) mark targets by killing tanks. Make-it-work modeling actions included setting probabilities of kill (PKs) to one. These actions are fine for demonstration purposes, but they can cause analysis biases and negative training. As always, there is tension between gaming the system and sticking to doctrine.

In part this tendency can be addressed by selling the value of distributed simulator training to pilots, who may perceive it as a threat to flying time. The critical point here is that simulator training is not "instead of" flying time--flying time is never going to be available in sufficient quantities. Instead, simulation can maximize the training benefits of precious flying time by giving pilots a feeling for large complex engagements, so they do not fly the live mission simply trying to gain situational awareness. Also, tactical ideas can be pre-screened so that live time is most efficiently used.

Scenario Discipline. Controlling such a large distributed scenario is difficult, a situation made worse by the lack of reliability of current ADS technologies. When a component crashes, how, when, and if its entities reenter the scenario must be under strict exercise control. This control is required to ensure scenario continuity (e.g., players not popping up in the middle of an engagement) and players returning with the proper situational awareness. Both of these difficulties were present in the STOW-E. Without these, battle outcomes may be very misleading. One approach to abate this difficulty would be to continually store player status (including situational awareness), so that conditions just prior to small interruptions could (sometimes) be sensibly recovered.

Another aspect of scenario discipline has to do with the practice of making certain entities invulnerable. This practice is sometimes motivated by the problems that would be caused by the loss of a high-value asset, such as an AWACS platform. However, such practices can cause mistraining in training applications, and at best will significantly complicate analysis based on such scenarios.

Entities should also not abuse "ground truth" information. This is almost standard practice for CGFs but was also observed for human participants in STOW-E. For example, at one cockpit simulator the pilot regularly looked at an auxiliary "stealth" display to assist him with locating ground targets. Again, this wreaks havoc with both training and analytic objectives.

Improvements Required for Stealths

Stealths (special displays) are used by instructors, exercise controllers, and analysts to passively observe distributed engagements. There are many features that we would like every stealth station to possess. The stealths we saw at STOW-E each possess some of these features, but others, especially those associated with access to voice communications and event-driven alarms (see below), were not present on any.

Viewpoints. One area of concern was that of available viewpoints, or perspectives. We, and others present, had great difficulty in understanding details of the unfolding STOW-E scenario. There was no one place where one could get both the macro view, such as overall positions, and the micro information, such as a platform's loadout and situational awareness, required for detailed understanding. As a result we had to split up and constantly visit different displays--and were still not satisfied with our ability to follow the battle. It would be an improvement if all displays allowed users to freely switch views between an overhead view, a view from the cockpit, or the viewpoint of an arbitrarily placed observer.

Finding particular platforms of interest, in a very large scenario, can be difficult with any kind of view. The ability to have designated platforms change color, or blink, would be extremely helpful, and appears easy to implement. Similarly, the ability to have "dead" platforms show up with a special symbology would often be valuable.

Extraction of Detailed Information. The small number of runs available implies that much of the analysis benefit will come from a detailed understanding of cause and effect relationships within a given run (battle). Therefore, improvements to stealth capabilities to provide the kind of detailed information analysts need should be of great value to ADS-based analysis efforts. Such improvements include the ability to designate an entity and have a pop-up window provide detailed information about the status of that entity. For computer-generated forces, this information should include current plans and intent.

Intent and plan information can be obtained for human participants by listening to their voice communications. We found that obtaining good situational awareness from a stealth was impossible without access to the communication that was taking place. It would be invaluable to be able to designate a player and then listen in on his comm. channel, or to be able to select an arbitrary comm. channel to listen in on.

Finally, the analyst will often be interested in particular types of interactions that occur relatively infrequently. It is fatiguing and error-prone to have to constantly watch a screen to determine when events of interest are occurring. Thus, one wants the capability to be alerted when a critical event like a theater ballistic missile (TBM) launch occurs, or when a strike mission is approaching its target.

Network/Site/Application Status. It would also be desirable for the stealths to display network, site, and application status. It was often difficult to tell when simulation entities went down. An automatic notification of such an event is essential to avoid confusion when monitoring in real-time.

Improvements Required for Network Reliability

For analysis, especially when there is a strong reliance on macro scenario outcomes like exchange ratios or win-loss decisions, system crashes can be catastrophic. Because few runs are typically available, each one is extremely valuable. However, including runs interrupted by network crashes, with the accompanying breakdown of remote site interactions, will almost certainly bias outcome measures of effectiveness (MOEs). For this reason nearly 100 percent network reliability is required for such analysis.

Application Gateways (AGs). The DIS protocols typically result in each site broadcasting PDUs that are essentially redundant. To help reduce the net traffic, applications gateways (AGs) were developed by the Naval Command, Control and Ocean Surveillance Center, RDT&E Division (NRaD) which filtered and compressed the wide area net (WAN) protocol data units (PDUs). The NRaD team put together and successfully implemented the application gateways in a very short time, with many sophisticated techniques (PDU culling, quiescent entity, data compression and bundling, etc.) substantially reducing bandwidth requirements. In fact, an impressive bandwidth reduction factor (on the order of 5:1) was achieved (see Tiernan et al. (1995) for details). Still, PDUs were lost. STOW-97's target of 50,000 entities is over an order of magnitude greater than that of STOW-E, so there is a lot more work to do. One AG problem that was observed was the loss of critical PDUs like fire and detonate. These unrecoverable PDUs must receive priority over self-correcting PDUs, such as entity position--which can be inferred.

DSI Reliability. STOW-E acted as a good testbed for the DSI software, and the Houston Associates staff felt confident that they understand and can easily fix many of the problems that did occur.

For the exercise, DSI reliability was good, with application reliability being more of a weak link. However, we found that the measure of network reliability used for STOW-E was misleading. This measure, based on connected site minutes, gave a 99.1 percent reliability value. A measure oriented toward the impact on the exercise from an analysis perspective would be preferable. Such a measure might be the percentage of time that *all* connectivity critical for analysis was present. This measure would have given a reliability figure of a little below 90 percent.

Improvements Required for DIS Maturity

Model Reliability. The number-one source of subsystem failure was the crashing of constructive components, such as (but by no means limited to) the MSIM model that was running at TACCSF. These crashes were likely caused by the models being stressed by large scenarios.[6] The only way to fully uncover these problems is by participating in such exercises. With larger exercises planned (e.g., STOW-97) the constructive elements will play an even more important role.

Unexpected PDU Types. Unexpected PDUs were a problem. This is not a short-term problem because unexpected PDU types will not go away: new and variant PDUs will be continually arising because of the need to look at new hardware, concepts of operation (CONOPS), and scenarios. Applications must thus be made capable of handling such PDUs gracefully.

Virtually every DIS analytic exercise will include experimental extensions to the nominal DIS standard. Obviously, extensions intended for *this* exercise are not "unexpected" but they could be from the perspective of a component for which they are not relevant, but which still "sees" them. For example, extensions to provide detailed radar jammer state information to potential jamees could be present in an exercise, and would be received by a simulation component such as a tank that does not include a radar. The tank simulator should not crash because of the receipt of such a PDU (it presumably should ignore the PDU).

Voice PDUs. The ability to embed communications into DIS PDUs is vital to make objects like stealths more informative, and, more important, to facilitate after-exercise analysis. However, under

---

[6]TACCSF points out that crashes to MSIM, in particular, were probably due instead to "unstable and untested interface software," as opposed to scenario loading. However, we are specifically aware of another simulation where scenario load was a problem, the Falcon Star F16 simulator at Grafenwoehr. In the case of the Falcon Star only a major effort during final testing averted serious overloading problems during STOW-E itself.

certain circumstances, such as limited bandwidth or the desire to exercise actual equipment, the communications cannot be embedded into DIS PDUs. Thus, this protocol should be optional.

Detonation PDUs. Detonation PDU improvements can resolve some of the terrain correlation problems. Clamping (artificially changing the altitude of an entity to make it appear to be resting on the surface of the earth) is commonly touted as a solution to minor terrain correlation problems. However, it can be shown that inconsistencies arise when munitions are employed against clamped entities and different applications control the munition and target. One way to correct this problem is to incorporate end-game geometry information into the detonation PDU, and require that the target use this geometry information--instead of munition and target EntityState--to evaluate damage.

Representation of RF Environment. Work on improving the representation of the radio frequency (RF) environment in DIS is under way.[7] Lack of these features could bias analysis results and result in negative training. We do not know whether or not Air Force requirements are being adequately addressed.

Data Logger Standard. Much of the data logger work we have seen provides only for inflexible "standard" reports to be generated. Analysis requires flexibility in examining data. Coupling the logged data to commercial off-the-shelf (COTS) database management systems (DBMSs) is unquestionably the most cost-effective way to accomplish this. In our experience, the cost of continually modifying home-grown report generators far exceeds that of acquiring and using a COTS DBMS. Fortunately, contracts have already been let to couple a data logger to such a DBMS.

Improvements Required for Complexity of Exercise Logistics

One feature that is apparent from participating in the STOW-E and BMDO TED is the enormous amount of effort (time and money) required to

---

[7]At the time of the STOW-E there were no electromagnetic PDUs. Thus, systems, such as radar warning receivers (RWRs), could not participate in the STOW.

simply get everything connected and nominally working. In the words of LTC J. D. Dennison, Air Force leader in the STOW-E: "The complexity of the environment (simulations, infrastructure, communications) necessary to maintain management/operational/technical control exceeds practicality." Little time, so far, is spent on anything beyond face checks that the disparate parts are interacting in sensible ways. This will need to change if analysis efforts in the DIS environment are going to be anything but occasional high-cost efforts. One item that would help is early scenario development and testing. Organized trouble reports might also help with identifying systematic errors.

The STOW-E provided an excellent opportunity to test and debug hardware and software. To obtain maximum benefits from such opportunities requires excellent trouble reporting and follow-up. There must be one place to record trouble reports and all participants must actively report. Additionally, where possible, trouble reports should be automated.

Several military participants noted that the ratio of contractors to military personnel was exceedingly high--10 to 1 was often flippantly quoted. To the extent that the contractor assistance might indicate "heroic" efforts to make STOW-E work, this ratio raises concerns about the cost and effort for practical training or analysis. Although certain dedicated facilities like TACCSF and the TBA normally have such high contractor-to-military ratios, this is not encouraging vis-à-vis broad and low-cost participation by many sites. We have seen no evidence that ADS is remotely close to the oft cited "vision" of a seamless battlefield which anybody can easily hook into.

Initially, the STOW-E was supposed to interact with the larger Atlantic Resolve exercise. That is, the results of a day of STOW-E combat would be fed into the Atlantic Resolve and vice versa. However, this did not occur. Indeed, all the Air Force missions were strictly scripted. It was often stated that this interaction should occur in future efforts. It is not clear to us who benefits from a direct linking. That is, whose training is improved? And, at what cost? These points are expanded upon in Section 4.

Improvements Required for Ability to Assess Training Benefit

Our study emphasis has not been on training, and we claim no special expertise in Air Force training. However, it seems quite clear that the training benefits of the STOW technologies need to be carefully considered. Dr. Herb Bell of Armstrong Laboratory, who can claim considerable expertise, indicates that AF/HRA has started to think about problems peculiar to training in a distributed environment.

Levels of Command. How many levels in the command hierarchy should surround the targeted trainees? We need to think explicitly about who is to be trained. The seamless battlefield concept is very nice, but how many levels of command (above and below) really need to be represented explicitly (and in particular with human-in-the-loop (HIL) participants) to train someone at a particular level? Does the joint task force commander really need a human flying an airplane around to be properly trained? Notice that very similar questions could also be asked in an analysis context. To begin answering this question we need to start by objectively quantifying the benefit of adding additional command levels.

Training Benefits. What are the training benefits of STOW-sized exercises versus other forms of (simulator) training? There are negatives to training in a distributed environment even if the environment is perfectly realistic and reliable. For one thing, training an operator may occur only once in a distributed exercise, as compared to many times in a stand alone context. In an air-to-ground mission, it would be possible to practice the weapons delivery phase over and over. In a distributed simulation, where this one mission occurs in a much larger context, the pilot would presumably have to fly the entire mission, and only get to do the weapon delivery phase once. It seems that both types of training are beneficial. Work will have to be done to optimize the overall training program.

Having the trainee embedded in a larger context also precludes certain standard training techniques such as stopping a run and providing immediate feedback, which is generally considered more beneficial than delayed feedback. We need to look for ways to compensate for this limitation.

A major benefit of distributed simulation is the potential it has to train entire teams, particularly geographically separated personnel. A partial capability, including giving team members the opportunity to work together in advance, and to become familiar with the terrain, should be available near term. Work will need to be done to develop means of evaluating the training benefits derived from such team training.

Well-Established Ways to Assess Individuals. There is a tendency in the ADS community to be excessively impressed by technology. When assessing the training benefits we must focus on well-established ways of measuring individuals. Realism may be sufficient, but it is not necessary for a good, or cost effective, training experience.

4. OBSERVATIONS FROM OTHER ACTIVITIES

To get a broad perspective we interfaced with a wide variety of ADS-related efforts, in addition to STOW-E. Included are exercises in which we have participated (BMDO TED, ABL Test Series 7), activities we are involved with (DIS Workshops, TBA Workshop), and exercises/activities we have followed (Warbreaker, A2ATD).

THE BMDO TECHNICAL ENGINEERING DEMONSTRATION

The technical engineering demonstration (TED) sponsored by the Ballistic Missile Defense Organization (BMDO) is a distributed effort run by the National Test Facility (NTF). The intent is to study several joint theater and national missile defense systems and architectures. At this point, the primary emphasis is to demonstrate ADS's potential for this type of evaluation. The TED is a joint effort with five distributed sites playing: the TBA, the NTF, TACCSF, RESA,[1] and the Aegis Computer Center at Dahlgren, Virginia. DIS protocols are used to connect the simulations at these sites.

The BMDO TED scenarios we participated in emphasized Air Force theater missile defense (TMD) and featured conceptual systems for ascent phase intercept and boost phase intercept (API/BPI). The TBA facility simulated a number of the components of this scenario using a mix of constructive and virtual entities, including theater ballistic missiles (TBMs), launch detection and tracking systems (Cobra Ball), an airborne laser (ABL), a kinetic kill missile launched from an F15-C (virtual simulator), an unmanned aerial vehicle (UAV), and an F15-E virtual simulator.

In addition to TBA, AFSAA played a role in the study, with participation by several AFSAA analysts. Much of this TED consisted of distributed wargames with very few, and thus immensely valuable, runs. To date, the emphasis has been on demonstrating the technical

---

[1]The Research, Evaluation, and Systems Analysis simulation at the Naval Command, Control and Ocean Surveillance Center (NCCOSC) RDT&E Division (NRaD). RESA is a constructive simulation of the naval warfare environment.

capabilities, so there has been little emphasis on detailed analysis questions such as the determination of engagement envelopes and their impact.

RAND, in its participation, tried to look beyond the immediate work of making the simulation run. Instead, we concentrated on providing the capabilities needed for an analysis that is capable of addressing the policy issues. In the broadest sense, this means we wanted to help make sure that the distributed simulation exercises provide some answers to questions concerning which supplemental TMD technologies are useful, and what command and control (C2) architecture changes are needed to make an augmented TMD system effective.

High-Level Observations of TED

The series of demonstrations illustrated some of the difficulties associated with distributed and interactive exercises for analysis. Key analysis issues include: The synthesized whole is very complex, there are very few runs available, output data are distributed and insufficient for many analysis applications, and post-processing tools are lacking.

The 15 December 1994 demonstration required a tremendous amount of effort just to coordinate the technical aspects of hooking nodes and simulations together. Little time seems to have been spent on what information was reliably obtainable from the runs. Validation relied mainly on face validity (i.e., does it look okay). Close examination revealed several problems that were not apparent from a face validity examination.

We believe that the paucity of runs requires wargame results to be supplemented with other methods to produce credible analysis. On one hand, constructive runs will be needed to determine engagement zones for concepts like API/BPI. On the other hand, virtual runs will be needed to calibrate critical timelines and other human factor assumptions in the constructive models. An attractive analysis plan for engagement zone evaluation would examine the constructive models for human factor assumptions, and then proceed to instrument the virtual runs in the TBA to capture data that could be used to calibrate the assumptions.

Finally, additional virtual runs near the engagement zone boundaries (as determined by constructive runs) could be made as a cross-check.

Problems Affecting BMDO TED Analysis

Some specific problems we encountered during our participation in the BMDO TED include:

Timestamp Problems. Participation in the TED revealed a number of problems that will become important when the BMDO effort goes beyond the demonstration phase. For example, a general lack of synchronization exists between simulation applications. As a result, timestamp information on PDU headers is meaningless. One of the ramifications of this is that EntityState PDU times are taken to be the time of receipt, for dead reckoning purposes. In a small simulation of an aircraft and an air-to-air missile interacting in the presence of latencies[2] we have shown that there is a significant impact on the engagements when timestamps are assigned with arrival times. In this experiment we simulated delays between the transmission of EntityState PDUs and their receipt, between the missile and aircraft components. Such delays would be typical of network latencies when the aircraft and missile are simulated at geographically distinct sites. Different ways of handling these delays, corresponding to common versus recommended practices, were examined. Details of this mini-study are presented in Appendix A.

Additionally, a lack of origin timestamps limits one's ability to perform useful post-exercise analysis with logged data. For example, the DIS protocols are structured so that it is not possible to directly associate "kills" with munitions detonations. This is not unreasonable, since in the case of multiple hits the assignment of who made the kill can be unclear.

DIS Protocol Problems. A number of DIS protocol problems revealed in the course of cursory data analysis (e.g., erroneous PDU data, missing PDUs, and entity removal problems) are minor and easily fixed, but they indicate the need for additional testing to ensure that

_____

[2]Latencies are the time intervals between the transmission and receipt of information between the components that make up the ADS simulation. The delays are due to processing and physical limitations--such as, the speed of light.

required PDUs are present and that all PDU data are correctly filled in. Generally, the testing for the December 1994 BMDO TED emphasized surface validity, that is, the testing sought to ensure that entity interactions appeared to take place correctly. Deeper looks, such as those suggested above, were generally neglected, so the fact that information was missing would not be uncovered unless the information was needed to make the entities interact, or appear to interact, correctly. This is not unreasonable for a pure demonstration, but if the objective is to demonstrate an analysis capability it is also important to ensure that the underlying information is complete enough to support post-run analysis.

Model Problems. Finally, a number of model problems have surfaced, such as models improperly having ground truth information. This is entirely expected, since many of the models were developed for demonstration purposes. However, in order to use them in an analysis context, assumptions and limitations must be known, and compared with fidelity requirements demanded by the analytic goals of the exercise. When models are inadequate they will need to be improved, or analysis strategies to work around their shortcomings will have to be developed. The limited number of runs in a fully distributed exercise reduces the opportunity to identify, correct, and test for potential limitations.

Most of the above difficulties can be attributed to the fact that ADS is new. However, since models and databases are constantly evolving, and the exercise managers will have limited opportunities to test the whole system, we can expect these difficulties to persist unless they are recognized as inherent problems that require the development of clear processes to minimize their impact. This places a real premium on after-action reports that detail problems, even if the problems were fixed--tracking such problems is an important first step toward the goal of addressing their basic causes.

Suggested Improvements Based on BMDO TED Experiences

Realistic Exercise Objectives. Getting credible analysis out of this type of exercise requires realistic objectives. It is not feasible, with a handful of DIS runs, to thoroughly compare multiple systems and architectures, or to assess engagement envelopes. When the fidelity of the models is low, analysis objectives are further limited. For instance, a comparison of different command architectures is not credible when no tracking or data fusion errors are present. On the other hand, a few such runs can be used to inform or calibrate other methods--such as constructive models--which can be run thousands of times.

Management of Exercises. The management of such exercises is extremely difficult, with joint participants at multiple sites. Some measures to enhance the quality of the exercises, though all at some cost, are:

1. Explicitly and rigorously test the components for adherence to DIS standards. Current testing is mostly by face validity (Do the simulated engagements appear to be realistic?) with many errors going undetected. For example, incomplete PDUs may have no effect at all on engagement outcomes if the missing fields do not affect the development of the engagements. However, their omission can stymie analysis efforts because needed data are not collected.

2. Routinely plan tests of the experimental setups. This can include pairwise testing among sites and pre-exercise test scenarios. Errors can then be identified and corrected prior to the main exercise.

3. Place a hold on software and database modifications at some point prior to the exercise. Last minute changes can induce errors that only show up during (or after!) the exercise. The limited time allotted to the exercise makes these errors difficult to correct.

4. Develop and use a set of predefined guidelines, i.e., a "game plan," for handling hardware and software failures. The value of one of the few analysis runs available can be greatly diminished by such a failure. When inevitable subcomponent failures occur, such as the loss of network connectivity or model crashes, exercise controllers must be able to quickly

recover, decide to restart the run, or otherwise account for the disruption.

Timestamps for All Air Force DIS Activities. As discussed above and in Appendix A, timestamps should be immediately added to all DIS PDUs. Time coordination among the sites can be most easily accomplished using GPS signals, but other alternatives, such as synchronizing over a voice telephone line, may be adequately accurate for many applications.

Summary Documentation of All Models. Many of the models used in DIS exercises were developed for demonstration purposes and have little or no documentation of their assumptions and limitations. Such documentation is invaluable for making preliminary assessments regarding the adequacy of an exercise configuration with respect to an analytic approach. Even such summary documentation would constitute a major improvement, although more comprehensive documentation will be important for efforts where the expertise is geographically distributed.

AIRBORNE LASER (ABL) TEST 7

Overview

The ABL Test 7 was a distributed analysis effort sponsored by the Airborne Laser System Program Office (SPO). The two participating sites were the TACCSF and TBA. Among the test objectives were to evaluate the ABL effectiveness in a variety of scenarios and to study potential retrograde methodologies on ABL survivability and impact on theater ballistic missile (TBM) defense. We focus on the study of retrograde methodologies for our discussion here. RAND's role was to review the analysis plan and design the experimental run matrix for the retrograde methodologies portion of the test.

Observations

In the retrograde methods portion of ABL Test 7, the virtual participants were an ABL pilot and mission commander at both the TACCSF and TBA. Constructive elements in the scenarios included threat TBMs, red fighters attacking the ABL (assumed to have penetrated the blue CAP), and automated laser weapon management. For these tests the human

ABL participants control their orbits so as to keep the plane safe while simultaneously performing their antiballistic missile mission. When attacked, these two are in conflict, and quite sensitive to human reactions and decisions. The key outcome measures of effectiveness were whether the ABL survived and how many TBMs were destroyed.

The input variables to be varied include the number of red attacking aircraft (one or two), two ABL rules of engagement, two ABL crews, red air keep-out range (three values), and the timing between the threat aircraft's run at the ABLs and TBM launches (three values). This exercise highlights the combinatorial difficulties associated with human-in-the-loop exercises. Running all combinations of the above variables requires 72 samples, not including samples to estimate variability or additional values for more inferences on the continuous variables. In addition, there were other variables the analysts would like to have varied but could not due to the limited sample sizes. The factors they believed could be critical included different threat aircraft, directions of attack, and TBM launch sequences.

The sample size constraints gave us an opportunity to use some advanced experimental design methods. (For more on these designs see Dewar et al., 1995.) The design used in the test is based on a 1/2 fraction of the 72 required samples for the full factorial design. This design is recommended in the National Bureau of Standards Applied Mathematics Series 58: "Fractional Factorial Designs for Experiments with Factors at Two and Three Levels," reprinted in McLean and Anderson (1984). We needed to use less than the full factorial design so we could fit the key effects and interactions into the allotted 40 runs and have replications--which are used to estimate the variance. This design is (nearly) orthogonal and allows estimation of all main effects (including quadratic for three-level factors) and all first order interactions, plus random error. Appendix B contains the case matrix.

The limited samples also suggest the use of constructive models to identify what factors are critical and the parameter ranges within which results may exhibit important sensitivities. In this example, some constructive pre-runs might have better focused on the red air keep-out range and the time interval between the threat aircraft's run at the

ABLs and TBM launches variables. This focus might have extracted more information from the scarce virtual runs. The analysis team had planned to do this if more time and resources had been available.

Another distributed analysis challenge that was evident in this exercise was reliability. The test exercise was postponed because display software could not keep up with the communications load, resulting in several crashes and unacceptable latencies. A subsequent exercise was successful. Analysts at the TACCSF, based on their years of experience in distributed exercises, recommend hardware, software, and scenario freeze dates so as to reduce the reliability risk.

To maximize the information taken from the human participants the analysts videotaped the participants from several angles. This way they could go back and record timelines and assess other human elements. This approach plays to the strength of the virtual simulations and is more objective and comprehensive than having the participants fill out after-action questionnaires.

WARBREAKER

ARPA's Warbreaker program, while not an Air Force effort, was one of the few distributed simulation efforts that has emphasized analysis-- and thus deserves special attention. This program to facilitate development of new concepts for theater missile defense includes the Warbreaker synthetic environment facility that used ADS to study theater missile defense options. Entities involved included humans in simulators and constructive models. During the time we followed Warbreaker activities, the simulation facility was heavily involved with the High Altitude and Endurance (HAE) Unmanned Aerial Vehicle (UAV) program, where it was used to further study the HAE system and to address performance issues through emulation. Their stated goals were (1) to get early user involvement in the program to develop, refine, and substantiate CONOPS; (2) to develop a generic HAE system emulation; (3) to analyze system design trades; (4) to support testing/operational demonstration planning and execution; and (5) to conduct assessment of future sensor, exploitation and C3 technology and system enhancements.

The Warbreaker program was way ahead of the community at large in utilizing synergistic interplay between virtual and constructive models. Much of the work we observed involved a UAV ground control station that included a mix of human operators and automated functions. Thoughtfully designed experiments were executed which involved alternating constructive modeling of this system (reacting to an influx of observations and requests for UAV services) with HIL experiments. The constructive models explored a range of possibilities and identified the more promising configurations for the ground control station. The HIL experiments checked these configurations using its more accurate operator representation, and measured human performance parameters that were used to calibrate the constructive models. Typical experiments were small in scale, involving at most a remote JSTARS simulator at Melbourne, Florida.

Warbreaker tools like SIMCORE may also be well in advance of other attempts at data capture for analysis in a distributed environment. In particular, SIMCORE "data probes" assist in capturing "why" something happened--these data are typically internal to simulation components--as well as "what" happened. It should be noted that Warbreaker exercises were relatively small in scope (in terms of sites and entities) when compared to either the BMDO TED or the STOW-E. The smaller scope facilitates connectivity, replication, and understanding issues. It also makes it a good place to obtain insights and develop analysis methods.

We should thus be able to learn much from Warbreaker. Accordingly, we are attempting to use some of their experience in our participation in other efforts.

DIS INTEROPERABILITY WORKING GROUPS

Part of our effort has involved participating in the semi-annual workshops on Standards for the Interoperability of Distributed Simulations. One motivation is to advance Air Force and analysis issues. Currently, the DIS standards are primarily oriented toward training applications, and have an Army flavor as well. This is a natural reflection of where the funding and early applications have come from. For instance, at the March 1995 Workshop fewer than 4 percent of the participants attended a meeting held for those with Air Force interest. We feel that the needs of analysis in general, and Air Force analysis in particular, require that the DIS standards include features that are not necessarily of interest to the majority of the participants in the DIS working groups.

COMPUTER-GENERATED FORCES (CGFs)

Computer generated forces is another area where improvement is needed. Situation awareness is very important for air combat analysis, but techniques to implement realistic situation awareness (SA) on the part of constructive participants need further development, and many CGFs do not even attempt to capture SA limitations. Simulations involving large numbers of aircraft will benefit from an ability to transfer control between models of lower and higher fidelity when engagements begin. This is a different type of transfer-of-control (TOC) than most DIS participants require. Finally, the ModSAF approach, which has the appearance of becoming a de facto standard, may be incapable of handling the extremely complex decisions needed for air-to-air combat in a few-on-few environment (see Kerchner, 1995), let alone the many-on-many environment which is typical of ADS scenarios. This issue needs to be examined carefully.

VV&A OF DISTRIBUTED SIMULATIONS

Analysis with distributed simulations is intimately connected with verification, validation, and accreditation (VV&A) of distributed simulations. This is a complex problem, and we want to help make sure that VV&A suitable for analysis, not only training, is addressed. Two very different approaches to validating distributed simulations are given in Dewar et al. (1995) and a Quality Research (1994) report.

CREDIBLE ANALYSIS SPECIAL INTEREST GROUPS

The community's growing concern about analysis issues is exemplified by the recent formation of a special interest group for credible analysis. Our participation allows us to interact with other researchers who are tackling analysis issues with DIS. The Institute for Defense Analysis (IDA), for example, has several interesting ideas regarding the storing of DIS generated data and subsequent analysis approaches. (See Stahl and Loughran, 1994, as well as other IDA publications listed in the bibliography.)

THE ANTI-ARMOR ADVANCED TECHNICAL DEMONSTRATION

Overview

One of the more ambitious early ADS analysis efforts is the Anti-armor Advanced Technical Demonstration (A2ATD). While we did not participate, the analysts involved in this Army effort briefed us on several of their lessons learned. In particular, they are addressing requirements for credible analysis with DIS. Among the original goals of the A2ATD was to "demonstrate a verified, validated, and accredited (VV&A) capability to support weapon system virtual prototyping, concept formulation, requirements definition, effectiveness evaluation, and mission area analysis on a combined arms battlefield at the Battalion or Brigade level."[3] The analysts also hoped to credibly estimate some anti-armor weapons' contribution to force effectiveness. This ongoing program consists of six diverse experiments on Army systems. Each experiment includes a mix of virtual (using upgraded SIMNET facilities) and constructive simulations. The virtual runs consist of both human participants and semi-automated forces (SAFORs).

Observations

The complexities and limitations inherent in large distributed analysis efforts have made some of the ambitious early goals unobtainable. Particularly difficult areas include overall complexity and reliability, limitations in the simulations' fidelities, small sample sizes, and primitive analysis tools. To combat this the analysts are using carefully designed experiments, including the randomization of human participants to virtual simulators. Additionally, the runs are carefully studied to obtain a detailed understanding of the mechanisms driving the simulation results. Much of the effort has concentrated on identifying and understanding differences between the virtual and

---

[3]Johnson (1994).

constructive simulations. Where appropriate, what is learned is being used to improve the models.

The iterative nature of the six experiments has facilitated a continuing improvement in DIS analysis capabilities, such as analysis tools. Through their experience with the A2ATD and other analysis efforts, the analysts at the Army Materiel Systems Analysis Activity (AMSAA) have developed "keys to credible (analysis using) DIS."[4] Some, but not all, notable keys are:

- DIS compliance is necessary but not sufficient for credibility.
- Problem definition and an evaluation plan are crucial in scoping a credible DIS experiment. Areas of particular importance lie in determining the questions to be answered, input data requirements, verification and validation required and feasible, and number of experiments that are required (or can be afforded?).
- Entrance criteria to determine readiness to execute a DIS experiment need to be established. Given the criteria, either do not start until they are satisfied or caveat results with respect to the deficiencies.
- Data must be certified and consistent across applications. This is particularly important for CGFs and environmental factors.

All of AMSAA's keys to credible analysis with DIS are consistent with our observations of other exercises. The emphasis must be on realistic objectives and time tables, and on designing experiments that recognize and account for limitations and uncertainties.

THEATER BATTLE ARENA

The Theater Battle Arena (TBA), an element of the Air Force Studies Analyses Agency (AFSAA), is located in the Pentagon, Washington, D.C. It was created less than two years ago to support the use of Advanced Distributed Simulation in support of AFSAA studies, and to highlight

---

[4]These were briefed to us by Will Brooks of AMSAA.

capabilities within the Air Force to support its warfighters through modeling and simulation (M&S). TBA has been involved with STOW-E, the BMDO Wargame, the ABL Test, and a half dozen other projects. Although we were also associated with a number of these projects, our involvement did not afford us the opportunity to interact extensively with the TBA during these exercises.

The TBA can play an important role as the Air Force becomes more committed to ADS projects. Because of the TBA's location in the Pentagon it can provide senior Air Force leaders a viewport to observe and interact with numerous ADS exercises. The TBA has also presented the Air Force Story to the public at numerous professional meetings such as those of the Air Force Association (AFA), the Armed Forces Communications and Electronics Association (AFCEA), and the Interservice Industry Training Systems and Education Conference (I/ITSEC). In our future study efforts we will become more involved with the TBA. Imbedded as it is in AFSAA, an organization with a strong analysis culture, the TBA may provide us with the best opportunity to observe and interact with projects focused on the uses of ADS for analysis.

5. OBSERVATIONS AND THOUGHTS ON USING ADS FOR ANALYSIS

In this section we present our conclusions regarding the advantages and challenges facing the users of ADS in general, and DIS in particular. We also discuss a broadly applicable approach to incorporating ADS into analysis, one that relies on an interplay of HIL experiments and constructive models.

ADS/DIS ADVANTAGES AND CHALLENGES

The advantages and challenges described in this section are relatively abstract. We compiled them by thinking about how the advantages and challenges we observed might be generally applicable to ADS analysis, and from first principles, such as the fact that HIL runs are usually constrained to real time. Of course much of the material originates from the work of others (see Bibliography); few items presented are entirely original with us. Illustrations of many of the advantages and challenges are provided in the discussions of ADS efforts, such as the STOW-E. The presentation in this section should help researchers determine when and how ADS may be of use to them.

Potential Advantages for Analysis with Distributed Models

Enthusiasm for ADS technology is partly based on its potential to address some of the primary shortcomings of most stand-alone constructive models--and thus to significantly enhance analysis efforts that currently rely solely on these models. These shortcomings include poor representations of human decisionmaking, processing constraints on entity and phenomenon resolution and fidelity, scenario extent, and the time required to extend a model or link models.

Better Treatment of Human Elements When Virtual/Live Participants Are Involved. ADS can potentially contribute to the reduction of all the above problems, but the most clear-cut advantage of the distributed interactive environment derives from its ability to incorporate a superior treatment of human characteristics, through the use of virtual (and live) participants. We found two noteworthy advantages associated with the inclusion of HIL into simulations used for analysis:

More Realistic Decisions. The appropriate inclusion of human players will generally provide more realistic decisions than when only constructive components are present. Behavioral areas where the realism associated with HIL is normally better than that of constructive decisionmakers include: (1) a sensitivity to situation awareness issues, (2) a propensity to become overloaded or stressed, (3) the ability to respond reasonably to unexpected situations, (4) the capacity to bring to the simulator years of warfighter experience in live systems. There is also the pragmatic advantage of the superior believability associated with HIL. Whether justified or not in a particular case, the ability of HIL to "sell" an analysis should not be underestimated.

Better Insight: Engagements Evolve in Ways Not Seen With Constructive Models. The impact of real human decisions (achieved with virtual and live participants) on the "flavor" of simulated air combat engagements is worth highlighting. We believe that it is vital to supplement statistical analysis outputs (exchange ratios, detection ranges, etc.) with the insight gained by examining the detailed interactions that take place in a small sample of engagements (runs). In a hardware trade study, for instance, this use of "plausible stories" provides an understanding of why hardware is helping, or failing to help. This understanding can lead to an analytical intuition about what features are most important; it can also inspire tactics changes that make better use of the hardware. These factors lead to a better assessment of the value of the system being studied, or of concepts for employing a system.[1] When warfighter participation is involved in the analysis, the flavor may be quite different (and presumably more realistic) than when only constructive entities are involved.

Parallel Processing Benefits. The massively parallel ADS environment facilitates the modeling of high resolution players and

---

[1]Of course, experienced pilots in virtual simulators should not be expected to "automatically" make optimal use of enhanced operational capabilities. For example, if they are initially exposed to a visual range missile with a very large field of view they should be expected to need a number of simulator trials to learn to take advantage of the enhanced capability, and then some more trials to integrate their new-found tactics into their general experience.

extensive scenarios, while simultaneously capturing a wider variety of phenomena than is practical to incorporate into a single model. Actual combat is extremely complex, with many factors that potentially affect battle outcomes. Thus, there is a constant pressure to increase the resolution and extent of most combat models. Examples include the desire to represent the environment (such as clouds and night) and joint theater-wide scenarios.

Faster Model Development. The use of a mature distributed system can potentially also speed analysis, since one would expect to have a large "library" of interoperable models that can be easily used together. To date, a significant library of such interoperable components has not yet been constructed. However, demonstration efforts, such as the BMDO TED, to interoperate DIS simulation components that have been developed for separate purposes strongly support (but do not per se prove) the notion that the effort needed to achieve analytic capability is far less than the effort needed to augment a stand-alone simulation to provide the same capabilities.

This is true even when there is an existing model with the required capabilities that can be incorporated--the process of incorporating one stand-alone model into another is typically very cumbersome.[2] However, one must be careful to distinguish between interoperability and compliance to the DIS standards. Compliance does not necessarily mean that the distributed models interact in a sensible way for the purposes of the analysis. Models developed by different people for varying purposes will undoubtedly have different assumptions and limitations. In an unpublished RAND study, R. A. Hillestad, J. Owen, and D. Blumenthal demonstrated that model outputs may vary considerably with the resolution of the modeled entities. Understanding the interactions and outputs of the linked disparate models will be an even greater challenge. It should be clear from this discussion that the use of ADS for analysis is tightly coupled to the DIS VV&A process.

---

[2]The problem stems not so much from anything intrinsic to stand-alone models as from the fact that because they are developed independently they rarely conform to a standard. The advantage of DIS here thus stems from the fact that the components conform to a standard that is intended to facilitate interoperability.

Challenges for Analysis with Distributed Environments

The potential analysis advantages of ADS are substantial. However, the nature of ADS poses significant challenges that are not as widely recognized in the community. Thus, we have centered our research on identifying these difficulties and the improvements required to overcome them.

Large Complex System Complicates Design and Analysis. Foremost among the challenges is the complexity of the synthesized whole, that is, the distributed combination of numerous constructive, virtual, and live entities. Ensuring that all the elements are at the right level of fidelity and will interact properly is a considerable challenge. We believe that the enormity of this difficulty is not fully appreciated by all in the ADS community, and that the complexity issue has not been fully addressed in the larger synthetic environments currently used. Community efforts to formally verify and validate distributed exercises will focus attention on this challenge.

Identifying and Accounting for Assumptions and Inaccuracies Are More Difficult. Within this complex synthesized whole it will be very difficult to identify, understand, and account for assumptions and limitations in the simulations and their interactions. The library of models concept previously mentioned carries with it the burden of determining and understanding limitations and inaccuracies in the simulation components being assembled for a study. This is especially true when geographic distribution is contemplated, since the expertise for the various components will be geographically distributed too! We do not believe that this problem can be solved simply by mechanisms such as video teleconferences--learning how to compensate for the lack of spontaneous interactions that take place when analysts and modelers work in close proximity will not be easy. Additionally, many inaccuracies will surface or pertain to the synthesized whole. Opportunities to identify and test for these will be limited to restrictions on the availability of the distributed sites and the network.

Distributed Output Data Increase Analysis Effort. Using multiple models will generally result in output distributed over multiple databases, greatly complicating analysis. DIS PDU loggers help a lot, but by themselves do not solve the analyst's problems: PDUs generally contain "what" information, while analysts also need "why" information. For example, an analyst might need to understand why one player did not engage another, but this information will be internal to that player's simulation application--and not contained in the PDU stream that application emitted. Tools such as the SIMCORE tool being developed for the Warbreaker effort attempt to address this kind of problem, and have some potential for broad use within the analytic community.

Runs Cannot Be Exactly Replicated. As described earlier, the ability to exactly replicate runs, at least for constructive components of a distributed application, is very useful to analysts. ADS, with latencies and human participants, will be unable to reproduce runs exactly. Reproducibility is vital when debugging software, especially when tracking down intermittent problems. At least as important is the help it provides in understanding unexpected phenomena that are observed.

Distinguishing Between New Real-World Insights and Model Artifacts Is Important. Tracking down the causes of unexpected phenomena is vital for good analysis, since unexpected phenomena may imply new insights about the real world scenario being studied. Of course, it is always necessary to ensure that one is not seeing the spurious result of a model artifact.[3]

Latency Introduces Inaccuracies and Complexities. When models are distributed, particularly over long-haul networks, the latencies introduced due to network effects add additional complexities that must be assessed and compensated for in the component simulations, or in the analysis itself. This will be of particular concern when fractions of a

---

[3]An interesting point can be made from the foregoing discussion. Notice that the focus is on tasks that the analyst must perform while going about his work. The technology can make some of them more difficult at the same time it makes others easier. In no case do any of these tasks disappear because of the introduction of ADS, or any other new technology. If we want to improve our ability to perform analysis we need to keep the *job* of the analyst clearly in mind, and avoid over-focusing on the *tools* used by the analyst. No new technology or model will, in the foreseeable future, do away with the need for analytic thought and planning throughout the analysis process.

second can alter outcomes--as in visual range air-to-air or tank-to-tank combat. These latencies will vary with network usage and thus add to the difficulties with exactly replicating simulation outcomes.

Management Is Complicated. The management of distributed analysis projects is unquestionably more complicated than that associated with the use of stand-alone models. We are concerned about configuration management when models and databases are updated, or when new versions of interfacing software are to be incorporated. Ensuring that updates are integrated into multiple sites and applications is a difficult problem, and one that has not been adequately addressed to date. Also, software and database versions vary from exercise to exercise. Typically, facilities like TACCSF and TBA are simultaneously involved in multiple exercises. Making sure that the proper software and data versions are used when switching between exercises is clearly a process that is both error prone and cumbersome, particularly in the absence of automated assistance. Tools such as common source databases that can input to multiple applications are part of the solution. Some fundamental research needs to be done on how to confirm, at exercise initialization, that the proper versions of software and data are being used by each simulation component.[4]

Scheduling Experiments Is More Difficult Across Multiple Sites. Another management problem is that of scheduling experiments that require the participation of multiple sites. This may be more complex for analysis than for training uses, because analysis typically proceeds in an iterative manner. There will often be unanticipated needs to conduct additional exercises. The constraints imposed by DSI network availability and the need to coordinate time and resources from many sites may severely limit the amount of time the synthesized whole is exercised. Thus there will be less time to test the system. However, adequate testing is clearly essential to the success of any analysis project, certainly for one involving ADS. Better organizational

---

[4]Since this was written, we have been exposed to an online service (Prodigy) that automatically detects obsolete versions of various tools and installs new versions during one's session. This is very much along the lines of the type of capability that we would like to see in an ADS environment.

relationships and improved network flexibility will be needed to allow the required testing time.

Reduced Reliability Is Inevitable. The end result of having to synchronize the models, databases, and environments of multiple distributed participants, while constrained by limited network availability, is reduced reliability. That is, it will be more difficult to find and correct for errors and limitations.

Challenges Are Introduced by Virtual/Live Participants. For the analyst, the use of human participants is probably the most exciting advantage of distributed simulation. However, analytic difficulties are introduced when humans are part of an experiment. The most obvious of these is the impossibility of achieving exact run replication when human participation is involved.

Human participants are also a challenge because they severely limit the number of runs that can be made for an experiment. Many analytic efforts require large numbers of runs to develop sufficient statistical power, and to adequately explore the parameter space. Indeed, experience has taught many analysts to insist on a minimum number of runs even with highly controllable models, such as JANUS. In typical applications RAND researchers find that about 30 runs at each scenario variation are required.[5] The number of scenario variants can be very large, so the total analysis can easily require thousands of runs. However, the number of runs possible with human participants will never be very large. For large distributed exercises there will be, at best, a few tens of trials total; smaller exercises might be able to obtain a few hundred runs. Methods of circumventing this sample size restriction will need to be developed for credible analysis. We believe that these methods will require a synergistic mix of virtual and constructive-only runs, with the latter probably being single site or even stand-alone. This idea will be expanded in Section 6.

_____

[5]The number 30 is a rough rule of thumb. One is really dealing with a signal-to-noise issue, where the "signal" is the effect being examined, and "noise" constitutes all of the other stochastic effects that obscure the effects of the "signal." The number of runs is determined by the "signal-to-noise ratio" implied by relative sizes of these influences.

Of course, for any analysis with human participants there are always concerns with individuals' learning curves and gaming the system. These factors directly confound attempts to assure the independence of separate runs, and clearly inject biases into simulation outcomes. Experimental design techniques, such as randomizing the participants to virtual simulators or formally incorporating learning effects, must be used to abate or account for these concerns.

Some Short-Term Technical Problems Exist. Finally, there are short-term technical problems, such as reliability and bandwidth. The current technology--hardware and software--is not reliable enough to handle hours-long exercises without some crashes. The maximum number of entities in an exercise was 1860 at the STOW-E. In this exercise, despite an application gateway compressing the net traffic by nearly a factor of 5, there were times the network could not handle the traffic. Future efforts plan on many more entities.

HOW TO USE ADS FOR ANALYSIS: A PROPOSAL

HIL Experiments and ADS

In this subsection we present an approach to incorporating ADS, and in particular HIL experiments, into analysis. This methodology did not originate with us. For instance, it is close to that practiced by the Warbreaker analysts, and similar ideas have been proposed in an unpublished RAND study by M. Callero, R. Steeb, and C. T. Veit. Some of the material in this subsection is directly adapted from an IEEE paper by Paul Davis (1995).

The most important way in which ADS experiments can inform analysis is by helping to ensure that human performance is adequately (for purposes of the analysis) represented in the analysis. This is important because analytical models and constructive simulations are unlikely to represent *critical* aspects of human performance well (e.g., whether pilots can assimilate and use a wealth of C^3I data quickly enough, under operationally realistic circumstances, to exploit proposed combinations of sensors, weapon systems, and operational concepts).

However, this does not imply that one can use ADS experiments involving HIL instead of experiments involving constructive runs. ADS experiments involving HIL are rarely adequate, by themselves, to meet the needs of the analysis. The most compelling (but by no means the only) reason for this inadequacy is that with rare exceptions the number of replications possible with HIL is limited to a few tens. Analysis requirements to explore a wide variety of conditions and to obtain good statistical accuracy can dictate the need for hundreds, and often many thousands, of runs. These constructive runs are needed, then, but if they are to have maximum credibility it is important that the constructive M&S representations of the critical human performance factors are compared with, and calibrated against, ADS runs involving HIL. ADS experiments can be designed to use virtual (or live) simulation to inform analysts regarding human factors, for instance to identify critical factors and ensure that they are in fact represented in the constructive models.

This point leads naturally into a second way in which ADS experiments can inform analysis. All experienced analysts realize that constructive models rarely incorporate all of the relevant qualitative factors that influence combat, despite ample evidence of their importance. This is likely to remain the case even when these factors are explicitly identified: While much progress has been made in the area of modeling qualitative factors (Davis, 1989; Dupuy, 1987), their incorporation into constructive M&S remains extremely difficult, and is unlikely to be adequate in the near term. One way to improve an analysis based on constructive models, by incorporating missing or poorly treated qualitative factors, is thus to directly utilize ADS runs with HIL participants.

Of course, since relying exclusively on HIL runs has just been shown to be unrealistic, the idea is to make the best use of the number of HIL runs available: Use constructive runs to identify key or critical cases, and explore these with ADS experiments utilizing HIL. This overall approach in which ADS HIL runs and constructive-only (possibly ADS) simulation mutually support each other is the essence of the concept.

Vision: DIS as Part of an Iterative and Cooperative Process Across Communities

Figure 1 provides a view of how to consider the emerging possibilities. In the center is the activity of designing, building, and *calibrating* constructive M&S. This activity, which is ongoing and highly iterative over a period of many years, often draws upon either small specialized DIS experiments (top right), as discussed previously, or even larger distributed war games (top left) for both insights and data. Further, this activity may inform the design of those experiments in the first place. The connection between the ovals "Conduct small, specialized ADS HIL exercises . . ." and "Analyze problems" is intended to refer to the direct contribution of ADS runs to analysis. As noted above, such analysis would normally be in conjunction with additional runs using constructive M&S. And, to make things more complicated but realistic, the M&S will typically be supporting and included in the

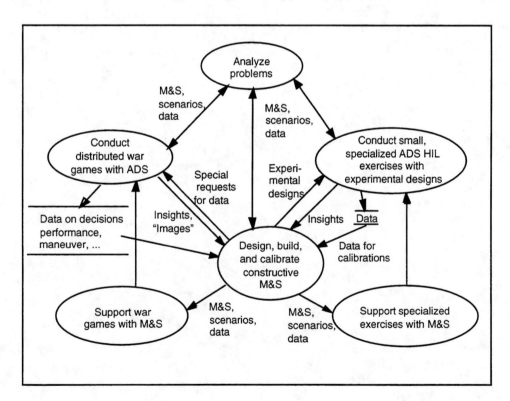

Figure 1--DIS-Mediated Interactions Among Model Building, Analysis, Experimentation, and Training

experiments and exercises. This is not a simple linear flow, but it is
perhaps the image of how we should view the continuing organic processes
of studying, innovating, experimenting, training, planning, and
analyzing if they are increasingly interwoven.

6. CONCLUSIONS AND RECOMMENDATIONS

STUDY CONCLUSIONS

Our more specific conclusions--that is, those that focus more specifically on Air Force issues, rather than broader ADS analysis issues--are set forth below.

ADS/DIS Has Potential To Supplement/Complement Other Analysis and Training Methods

We see a large potential role for ADS/DIS in Air Force analysis, training, and mission rehearsal. For analysis, the ADS concept addresses some of the critical weaknesses of traditional analyses supported by stand-alone models.

Many Improvements Are Required To Realize Air Force Potential

However, as emphasized throughout, many improvements are needed to realize this potential. The improvements required for the Air Force differ from those for the primary developers (U.S. Army and ARPA) of ADS technologies. To ensure that the Air Force gets the most from ADS technologies an Air Force investment strategy for ADS is needed.

The Current Focus Is on Demonstrating Technology

The current focus of most ADS exercises is on demonstrating and testing the technology. For training and analysis the costs in time and money are currently excessive. For ADS to become a practical, efficient tool for analysis or training, the excessive costs and complexity must be reduced. As we have noted, for ADS to effectively support an analysis effort, it must be part of a wider research strategy, probably one making extensive use of traditional models and methods. As with any constrained resource, the high cost of ADS will reduce the utility it provides to analysis. It will be some time before widespread analytic uses are credible and cost effective.

To some extent, the extra cost of ADS, versus stand-alone modeling, may be intrinsic; it may be a price one pays for the additional capability that goes along with ADS. This is, in our opinion, an open question. There is clearly a large up-front investment needed in ADS, but for something like the synthetic battlefield vision (whether the reader likes it or not is not the issue here), it is plausible that the infrastructure maintenance costs, spread over many applications, will be quite modest.

If this overhead should indeed be modest, the issue "Is ADS worth it for my analysis?" would then be largely one of comparing the ADS costs specifically associated with an analysis effort to those associated with alternatives. The alternatives might be (1) attempting to obtain the required behavioral realism from constructive models, or (2) scaling down the goals of the analysis (which might entail additional costs to the supporting program due to a resulting inferior decision made on the basis of the analysis). In the first case, given the difficulties of simulating human decisions, it is plausible that the modeling enhancements needed would sometimes exceed that of "renting" ADS facilities, including HIL participants. In the second case, program costs such as extra modification cycles, or worse, building the wrong system to implement, could easily exceed the costs of using ADS, even after these costs are scaled down by a factor equal to the probability that a wrong decision is made due to the omission of ADS from the analysis.

The above sort of trade has not, to our knowledge, been addressed. It is even more complicated to perform than the above issues indicate, because additional factors, such as the time available to do the analysis and the capabilities of the staff, also need to be considered.

ADS Has Potential for Air Force Training and Rehearsal

A potential near-term benefit from ADS is in training participants other than pilots[1] and in rehearsing missions. AWACS and JSTARS crews, for instance, do not require the out-of-cockpit visual fidelity that pilots do, and thus can get the experiential benefits of command, control, and communications using workstation displays. For mission

---

[1]This is not to say that pilots cannot benefit too. It is to emphasize that the skills being trained are those related to human interactions--as opposed to flying skills.

rehearsal, combatants, such as pilots, can run practice missions to familiarize themselves with the terrain, defenses, and other, perhaps geographically remote, mission elements. *In each of these, the training focuses on cognitive features of the human participants, rather than the quest for absolute realism in simulators and models.*

RECOMMENDATIONS

Air Force Should Develop ADS Investment Strategies

The great potential benefits that ADS can bring to the Air Force, coupled with the large list of needed (but non-trivial) improvements, imply that the Air Force should invest in ADS, but that a thoughtful process is needed to make the investments successful. Our main recommendation is that the Air Force develop a comprehensive strategy to implement ADS improvements--much like the Army's Master Plan (Department of the Army, 1994). This strategy will facilitate moving beyond the demonstration stage of ADS usage, and into regular utilization for analysis and training purposes. The investment strategy will have to balance the potential benefits versus the costs and technological risks. A good starting point would be to prioritize the improvements required-- such as those listed earlier in this report.

Analysis and Training Strategies Should Be Emphasized in Developing Future ADS Plans

For analysis efforts using ADS, specific research strategies must be developed that focus on the advantages ADS provides while working around the problems, including the fact that ADS constitutes a highly constrained resource. This is a complex issue: the analysis strategy cannot be developed ad hoc after the runs have been made. In particular, the need for an ADS component in an analysis should be driven by the needs of the analysis team not simply by the fact that the component is available.

Automated Tools and Procedures That Assist in Managing ADS Efforts Should Be Aggressively Developed

We have noted that the effort spent on ADS exercises is great. It is also rather error-prone, as is to be expected for such a new technology which requires many individuals to cooperate in a complex and nonroutine environment. Automated tools for specific tasks, and standardized procedures (including both manual and automated steps) can reduce both effort and errors. One area where automated tools would be extremely beneficial is in assisting the distribution of databases and software upgrades to distributed components. Such tools would function both to assist in the actual distribution and installation, and to ensure (at exercise initialization) that the proper versions are in fact being used.

Another area that is a candidate for procedure development involves testing the behavior of CGF and other simulation components. Such testing does not appear to be a good candidate for full automation, because this validation phase will generally have unique features for each exercise. However, procedures and partial automation should be helpful.

These procedures and automated tools are not intrinsically Air Force specific, although the validation-oriented tools will likely benefit from testing oriented toward Air Force systems and missions. However, we believe that the DIS community as a whole has been somewhat slow to recognize that many awkwardnesses encountered in current DIS exercises are not one-time occurrences that one just works through. Rather, they are symptomatic of systemic problems likely to recur again and again, but they are amenable to solutions. Someone needs to step up to the challenge of mitigating these problems; if undertaken by the Air Force, then Air Force interests will more assuredly be considered.

Recommendations That Can Be Implemented for Each Exercise

The following recommendations can be immediately implemented for all exercises that include analysis objectives:

Define Realistic Exercise Objectives. Getting credible analysis out of this type of exercise requires realistic objectives. It is not feasible, with a handful of DIS runs, to thoroughly compare multiple systems and architectures, or to assess engagement envelopes. When the fidelity of the models is low, analysis objectives are further limited. For instance, a comparison of different command architectures is not credible when no tracking or data fusion errors are present. On the other hand, a few such runs can be used to inform or calibrate other methods--such as constructive models--which can be run thousands of times.

Improve Management of Exercises. The management of ADS exercises is extremely difficult, with joint participants at multiple sites. Some measures that would enhance the quality of the exercises, though all at some cost, are:

- Explicitly and rigorously test the components for adherence to DIS standards.
- Routinely plan tests of the experimental setups. This can include pairwise testing among sites and pre-exercise test scenarios. Errors can then be identified and corrected prior to the main exercise.
- Place a hold on software and database modifications at some point prior to the exercise.

Implement Timestamps for All Air Force DIS Activities. Timestamps should be immediately added to all DIS PDUs. Time coordination among the sites can most easily be accomplished using GPS signals, but other alternatives, such as synchronizing over a voice telephone line, may be adequately accurate for many applications.

Develop Summary Documentation of All Models. Many models used in DIS exercises were developed for demonstration purposes and have little or no documentation of their assumptions and limitations. Such documentation is invaluable for making preliminary assessments about the adequacy of an exercise configuration with respect to an analytic approach. Even such summary documentation would constitute a major improvement, although more comprehensive documentation will be very important for efforts where the expertise is geographically distributed.

FUTURE RESEARCH DIRECTIONS

Finally, we describe two major efforts for our project as it continues into FY1996. These will lead to a plan for Air Force investment in ADS technologies, and an analyst's guide for utilizing ADS.

Air Force Investment Plan for ADS

A large gap exists between current ADS capabilities and those needed to support Air Force needs in analysis and other areas. It is unrealistic to pursue all shortcomings at once. Selecting which improvements to pursue will require a clear understanding of how they benefit the Air Force. This understanding, needed for efficient investment, can best be obtained by developing and executing a top-down requirements-driven investment plan as shown in Figure 2. The investment plan must consider existing systems, planned programs, and potential concepts. We would like to start by building a framework for evaluating ADS enhancements, and then perform an initial assessment of an investment plan using this framework. A more comprehensive effort would be performed as a follow-on effort.

Note that a benefit of a requirements-driven investment plan is that it entails distinguishing between different categories (i.e. purposes) of analysis, which can in fact be expected to be affected by ADS in different ways. An incomplete list of such categories or

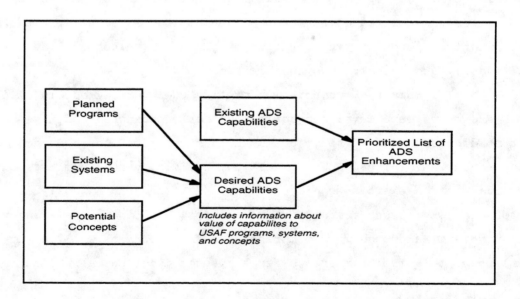

Figure 2--Air Force Investment Plan for ADS

purposes includes concept evaluation for new systems, performance analysis for system development, requirements analysis, deficiency identification, force structure analysis, and joint doctrine development. The differing nature of analysis for different purposes has been pointed out by several sources, including Dewar et al. (1995), and by Phil Thayer of ACC/XP-SAS in private correspondence.

An additional benefit of a requirements-driven investment plan is that it contributes to the goal of *quantitatively* evaluating the benefit of ADS to various Air Force programs. As noted earlier, a quantitative evaluation of that nature does not appear to have been undertaken.

ADS Analyst's Guide

To make our findings most useful to the analytic community, we intend to publish an "ADS Analyst's Guide" that will assist those who want to determine if ADS should play a role in their analysis. When ADS becomes a component of their study, the Guide will assist analysts in all phases of their effort, including exercise and experimental design, exercise execution, and analysis. If it fully meets its objectives, the Guide will provide both positive and negative support to analysts. Positive support includes broad analytic strategies that best utilize different tools, such as virtual (human-in-the-loop) ADS exercises and constructive models. The models may or may not take the form of a distributed simulation. "Negative" support includes warnings of problems that can be anticipated and techniques to mitigate them.

We are developing an approach for credibly using ADS for analysis which includes both the application of back-to-basics scientific principles and lessons learned from the investigation of ADS analysis-oriented efforts. Designing the experimental synergy between classical methods and the new technologies is the backbone of this approach. Generally, we look at ADS exercises as part of a broader research strategy. To overcome the limitations of making critical inferences based only on a relatively small number of ADS runs, we recommend that the ADS runs be preceded by preliminary analysis based on a set of constructive model runs to determine which scenarios and factors are most important. Afterward, additional constructive runs, made more credible by ADS runs with human participants, can fill out the experimental matrix.

This approach will be used and documented in an actual ADS effort involving analysis. The selection of the project is critical; it will be coordinated with XOM and AFSAA. Timing is another important factor: the main ADS exercise should be scheduled toward the end of the year, since several time-consuming steps must precede it. These include the preliminary analysis using constructive simulations. The analysis, along with sophisticated design of experiments, will be used to carefully construct the ADS cases to be run. The outputs of the various runs will be analyzed and the results synthesized. These experiences, supplemented by lessons learned by others, and by our experiences with STOW-E and the BMDO TED, will form the foundation of an ADS Analyst's Guide, describing how to use ADS for analysis. Given time and budget constraints, many areas of the Guide will not be fully developed this year. In fact, we expect the Guide to be a dynamic document that is routinely augmented and revised as the technology of ADS analysis evolves.

Appendix A
THE IMPACT OF NETWORK LATENCIES

In the December 1994 BMDO TED, PDU header times were generally not used, or if used, were useless because no attempt was made to synchronize clocks. On receipt of a PDU, any times associated with it were assumed to be the time of receipt. This has serious consequences. For instance, in the case of EntityState PDUs, the states could be out of synchronization by as much as 300 msec (max latency allowed by DIS protocols). A missile guiding on a target traveling at 1000 ft/sec (roughly mach 1) but using target EntityState PDUs lagging by 300 msec might pass behind the target by as much as 300 feet instead of impacting it. It is true that the application associated with the missile would declare a hit, but the application associated with the target would evaluate the kill, and it would not agree that the missile came close.

In general, the belief that it is acceptable to use arrival timestamps, because things appear consistent in each application, is fallacious. The most obvious way in which inconsistencies can arise occurs because network delays are not constant. Consider an entity moving with constant speed S. It will appear to trail its true position by an amount $S * dT$, where dT is the delay for the message. However, a change in the delay between two EntityState PDUs for an entity, from dT_1 to dT_2, will induce an apparent jump in the position of the entity of magnitude $S *(dT_2 - dT_1)$. For an aircraft traveling at 1000 ft/sec, roughly mach 1, a change in the delay of 50 msec, not impossible, translates to a 50-foot jump! Imagine trying to fly close formation with this aircraft!

However, even if the delays were constant, significant, albeit more subtle, problems would exist.

The plots in Figures A.1–A.3 were derived from a simple simulation in which a "missile" attempted to impact a "target" based on target information received in EntityState PDUs. The simulation had the capability to simulate network latencies, and alternative ways in which the receiving applications projected states based on received EntityState PDUs.

The plots show the "true" separation of missile and target. The sharp dips in range correspond to closest approach passes of the missiles. In Figure A.1 no latency was present; EntityState PDUs were up to date. In Figure A.2, a 300 msec delay was present, but correct timestamps were used. The guidance information used by the missile might thus be slightly "stale," but there was no systematic time shift. The range plots are essentially identical in these two cases.

In Figure A.3 the time associated with the target state was taken to be the time of the PDU's arrival, and was thus 300 msec off. Here, we see that the range distribution is significantly different, and in fact the actual range to target was never less than 35 feet, versus 4 feet when correct timestamps were used. Errors were not as extreme as the 300-foot value suggested earlier because the engagement crossing angles were not the most extreme possible.

An additional problem arises when arrival times are used to label the states of incoming EntityState PDUs. Network latency is not an invariant, but will generally vary with network load and other factors (the speed of light is generally a small part of the total latency, unless geosynchronous satellite links are involved). Varying message delays will have the effect of causing entities to appear to jump by distances equal to the product of entity speed and change in delay. For a fast-moving entity, even smoothing algorithms in the dead reckoners will not resolve the situation.

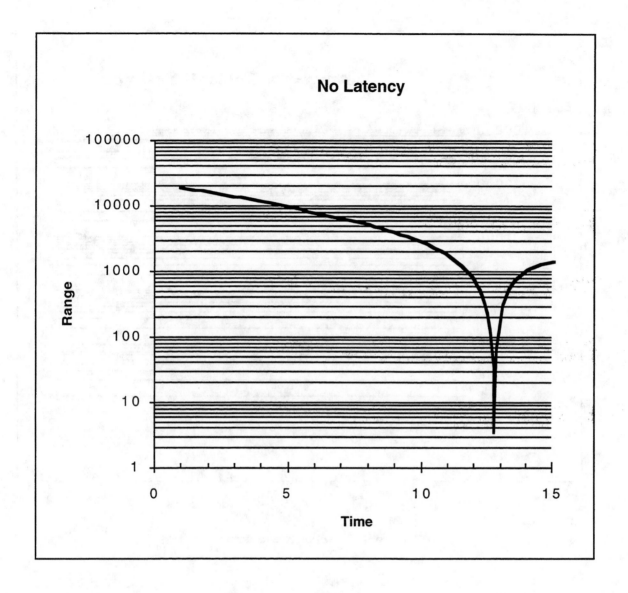

Figure A.1--No Latency

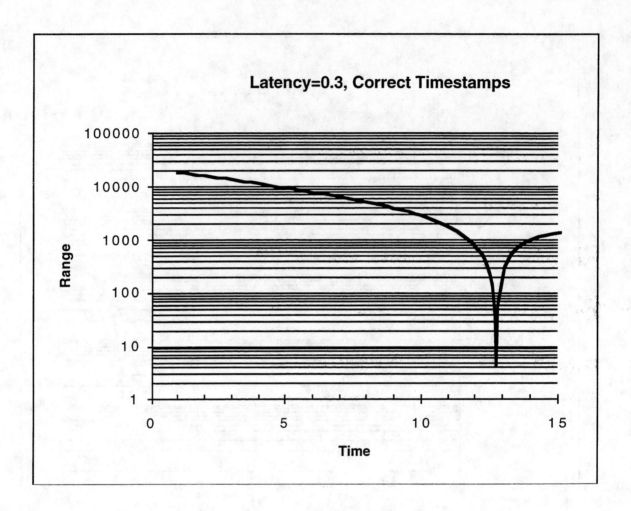

Figure A.2--PDU Timestamps Used

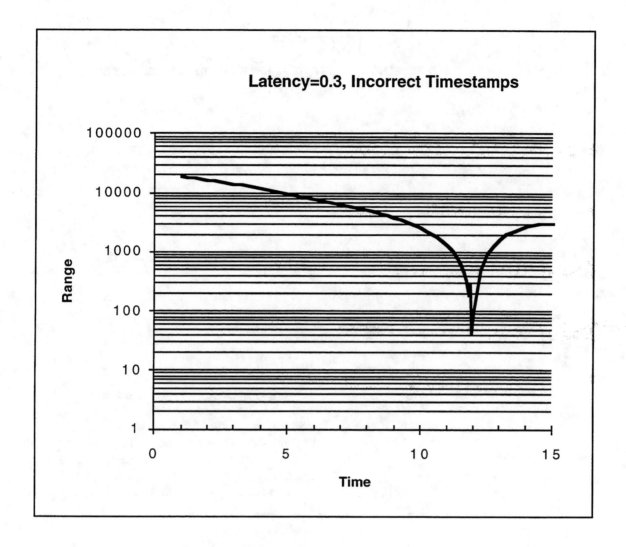

Figure A.3--Arrival Timestamps

Appendix B

ABL CASE MATRIX FOR VIRTUAL RETROGRADE EXPERIMENTS

Listed in Table B.1 are the factors that were varied in the Airborne Laser Test Series 7 Retrograde Study. Table B.2 presents the test case matrix.

Table B.1

Notation for Variables

Variable #	Label	Values	Description
(1)	AC	1,2	# threat Aircraft (1,2)
(2)	ROE	0,1	ROE against soft targets. Shoot(=1) or don't shoot(=0) at red air
(3)	CREW	1,2	Two sets of crews (pilot and mission commander).
(4)	TIME	(-x,0,x)	Denotes relative (to a nominal value) between red air run at ABL and TBM launch start
(5)	RANGE	(y1,y2,y3)	Keep out range that ABL will react to inbound red.

Table B.2

Case Matrix

Run	AC	ROE	Crew	Time	Range
1	1	0	1	-x	y1
2	1	1	2	-x	y1
3	2	0	2	-x	y1
4	2	1	1	-x	y1
5	1	0	1	0	y3
6	1	1	2	0	y3
7	2	0	2	0	y3
8	2	1	1	0	y3
9	1	0	1	x	y2
10	1	1	2	x	y2
11	2	0	2	x	y2
12	2	1	1	x	y2
13	1	0	2	-x	y2
14	1	1	1	-x	y2
15	2	0	1	-x	y2
16	2	1	2	-x	y2
17	1	0	2	0	y1
18	1	1	1	0	y1
19	2	0	1	0	y1
20	2	1	2	0	y1
21	1	0	2	x	y3
22	1	1	1	x	y3
23	2	0	1	x	y3
24	2	1	2	x	y3
25	1	0	2	-x	y3
26	1	1	1	-x	y3
27	2	0	1	-x	y3
28	2	1	2	-x	y3
29	1	0	2	x	y1
30	1	1	1	x	y1
31	2	0	1	x	y1
32	2	1	2	x	y1
33	1	0	2	0	y2
34	1	1	1	0	y2
35	2	0	1	0	y2
36	2	1	2	0	y2
37	2	0	1	-x	y2
38	2	0	2	-x	y1
39	2	1	1	-x	y1
40	2	1	2	-x	y2

BIBLIOGRAPHY

Christenson, W., and R. Zirkle, "73 Eastings Battle Replication--A JANUS Combat Simulation," IDA Paper P-2770, September 1992.

Davis, P. K., "Modeling of Soft Factors in the RAND Strategy Assessment System (RSAS)," RAND, in MORS, Alexandria, VA: Military Operations Research Society, *Mini-Symposium Proceedings: Human Behavior and Performance as Essential Ingredients in Realistic Modeling of Combat— MORIMOC II*, 1989; also Santa Monica, CA: RAND, P-7538, 1989.

Davis, P. K., "Distributed Interactive Simulation in the Evolution of DoD Warfare Modeling and Simulation," *Proceedings of the IEEE*, Vol. 83, No. 8, August 1995.

Defense Science Board Task Force on Simulation, Readiness and Prototyping, *Impact of Advanced Distributed Simulation on Readiness, Training, and Prototyping*, Office of the Under Secretary of Defense for Acquisition, Washington, D.C., January 1993.

Department of the Air Force, *A New Vector*, June 1995.

Department of the Army, *Distributed Interactive Simulation (DIS) Master Plan*, September 1994.

Dewar, J. A., S. Bankes, J. Hodges, T. Lucas, and P. Vye, *Credible Uses of the DIS environment*, Santa Monica, CA: RAND, MR-607-A, 1995.

DIS Steering Committee, *The DIS Vision: A Map to the Future of Distributed Simulation*, Institute for Simulation and Training, University of Central Florida, October 1993.

Dupuy, T. N., *Understanding War*, New York: Paragon House Publishers, 1987.

Johnson, I. M., "Anti-armor Advanced Technology Demonstration Experiments 2,3,4, and 5 Independent Evaluation Plan and Test Design Plan," AMSAA, Combat Integration Note No. DN-CI-1, May 1994.

Kerchner, R., "Scaling Problems Associated with Rule-Based Decision Algorithms in Multiple-Objective Situations--Value-Driven Methods as an Alternative" (in manuscript), 1995.

McLean, R. A., and V. L. Anderson, *Applied Factorial and Fractional Designs*, Marcel Dekker, 1984.

Quality Research, *Methodology Handbook for Verification, Validation, and Accreditation (VV&A) of Distributed Interactive Simulation (DIS)*, Institute for Simulation and Training, University of Central Florida, September 1994.

Stahl, M., and J. Loughran, *A Repository for DIS Data and Tools,* Position Paper 94-120 presented at the 11th DIS Workshop, September 1994.

Stahl, M., and R. Schwartz, "Using SIMNET for Military Analysis," draft IDA Paper, October 1993.

Schwartz, R., and D. DeRiggi, "SIMNET-based Tests of Antihelicopter Mines," IDA Paper P-2913, January 1994.

Tiernan, T. R., K. Boner, C. Keune, and D. Coppock, *Synthetic Theater of War-Europe (STOW-E): Technical Analysis*, Naval Command, Control and Ocean Surveillance Center, May 1995.

TRADOC Analysis Center, White Sands Missile Range, *Results of the M1A2 SIMNET-D Synthetic Environment Post-Experiment Analysis*, May 1993.